Return to the Sea

Return to the Sea

The Life and Evolutionary
Times of Marine Mammals

Annalisa Berta

Illustrated by
James L. Sumich and Carl Buell

UNIVERSITY OF CALIFORNIA PRESS

Berkeley   Los Angeles   London

University of California Press, one of the most distinguished university presses in the United States, enriches lives around the world by advancing scholarship in the humanities, social sciences, and natural sciences. Its activities are supported by the UC Press Foundation and by philanthropic contributions from individuals and institutions. For more information, visit www.ucpress.edu.

University of California Press
Berkeley and Los Angeles, California

University of California Press, Ltd.
London, England

Library of Congress Cataloging-in-Publication Data

Berta, Annalisa.
  Return to the sea : the life and evolutionary times of marine mammals / Annalisa Berta ; illustrated by James L. Sumich and Carl Buell. – 1st ed.
      p.   cm.
  Includes bibliographical references and index.
  ISBN 978-0-520-27057-2 (hardback)
    1. Marine mammals—Evolution.   I. Title.
QL713.2B475   2012
599.5'138—dc23   2011052056

19   18   17   16   15   14   13   12
10   9   8   7   6   5   4   3   2   1

Cover image: Representative marine-mammal community during the Pleistocene (clockwise from top left): sea otter (*Enhydra lutris*), Risso's dolphin (*Grampus griseus*), polar bear (*Ursus maritimus*), northern fur seal (*Callorhinus ursinus*), dugong (*Dugong dugon*), walrus (*Odobenus rosmarus*), and bowhead (*Balaena mysticetus*). Painting by Carl Buell.

For my academic children:
Sharon, Peter, Amanda, Carrie, Liliana, Rocky, Mandy,
Fran, Megan, Lisa, Alex, Morgan, Rachel, Josh, Breda,
Cassie, Celia, Samantha, Sarah, Jessica, Will, and Nick.

The past is never dead. It's not even past.

—William Faulkner, *Requiem for a Nun*, 1951

CONTENTS

PREFACE

This book grew out of my thirty years of teaching in the Biology Department at San Diego State University. Although I have mostly taught biology majors and graduate students it was the challenge of teaching non–science majors that really brought home the importance and need to effectively communicate science to the public. Teaching a course for nonmajors, I learned to eliminate the jargon, emphasize concepts, and provide rich, well-illustrated examples to clarify major points, an approach I have attempted to follow here.

My goal in writing this book is to use marine mammals—their enormous appeal and charisma—as a vehicle to present aspects of their diversity, evolution, and biology and, more generally, science and scientific thought. Accordingly, I present various controversies, test alternate hypotheses of explanation, and evaluate and interpret the available evidence.

As an evolutionary biologist, I focus on the role that evolution has played in the marine mammals we see today. It is the thread of evolution and knowledge of the history of these fascinating mammals that helps us to understand their present-day diversity and responses to environmental challenges. A historical evolutionary framework for marine mammals set against a backdrop of changing climates and geography offers a valuable perspective and, in many cases, lessons for the future. I discuss what we know as well as how we know about the diversity, evolution, and biology of marine mammals. I also inform readers about the patterns of change that are taking place today, such as food webs and predator–prey relationships, habitat degradation, global warming, and the effects of humans on marine mammal communities. The future of marine mammals depends on each of us—scientists as well as the informed public—working together to avoid crises before they develop or to appropriately manage those that arise. In the words of the novelist William Faulkner, "the past is never dead. It's not even past." In this context, the heritage of extinct marine mammals is a key element embodied in the life and evolutionary times of living marine mammals.

ACKNOWLEDGMENTS

I especially appreciate the efforts of my talented collaborators Jim Sumich, who provided most of the excellent line drawings and photographs, and Carl Buell, for his exquisite paintings of extant and extinct marine mammals. The many other colleagues who contributed photographs and line drawings are identified in the captions and in the credits at the back of the book. Valuable comments came from students and colleagues, especially current graduate students in my lab: Sarah Kienle, Jessica Martin, and Samantha Young. I also thank John Gatesy, Jonathan Geisler, and Hans Thewissen for granting permission to use art that was originally commissioned by Carl Buell for their research. The editorial staff at University of California Press, including Lynn Meinhardt, editorial coordinator; Kate Hoffman, project editor; Jason Hughes, project manager; Rachel McGrath, editorial manager; and Chuck Crumly, publisher, are gratefully acknowledged for their assistance in preparation of this book.

CHAPTER ONE

Marine Mammals

An Introduction

Mammals, like nearly all other tetrapods (or four-legged animals), evolved on land. Marine mammals are a diverse assemblage of at least seven distinct evolutionary lineages of mammals that independently returned to the sea and include whales, dolphins, and porpoises (**Cetartiodactylans**); seals, sea lions, and walruses (**Pinnipedia**); sea cows (**Sirenia**); extinct sea cow relatives (**Desmostylians**); polar bears; sea and marine otters; and extinct aquatic sloths. The secondary adaptation of mammals to life in water required various morphological specializations, including for some lineages dramatic changes in body size and shape compared to their terrestrial relatives. Marine mammals are relatively large, with streamlined bodies and reduced appendages (for example, small or no external ears) and thick fur or fat layers for insulation. Other modifications for swimming and diving include the transformation of limbs into flippers and/or use of the tail for propulsion in water.

The story of marine mammal diversity, evolution, and adaptation is intriguing. Where they originated and how they evolved provides a historical framework for understanding how marine mammals make a living today, guiding our future efforts in their conservation. Before telling this story, I need to introduce some basic information about the various groups of marine mammals.

MAJOR GROUPS OF MARINE MAMMALS

Marine mammals include approximately 125 **extant** (or currently living) species that are primarily ocean dwelling or dependent on the ocean for food. The polar bear, while not completely aquatic, is usually considered a marine mammal because it lives on sea ice most of the year. Fig. 1.1 shows the major groups of marine mammals and the numbers of living species. Marine mammals range in size from a sea otter, weighing as little as 1 kg (2.2 lb) at birth, to a female blue whale, the largest mammal to have ever lived, weighing over 100,000 kg (2,200 lb). Marine mammals live in diverse aquatic habitats around the world, including salt, brackish, and fresh water, occupying rivers, coastal shores, and the open ocean.

Apart from diversity in size and habitat, marine mammals are fascinating in a number of respects further explored in this book. Most are capable of prolonged and deep dives on a single breath of air. Such extreme diving requires a remarkable suite of anatomical and physiological specializations. Some whales undertake long annual migrations, among the longest known for any animal. Most feed on fish and various invertebrates, such as squid, mollusks, and crustaceans. Some whales filter water and prey through uniquely developed sieves, baleen plates, that hang down from their upper jaws. The remarkable ability to produce and receive high-frequency sounds among other whales has allowed them impressive navigation skills and the ability to precisely locate prey underwater. A few marine mammals, the sirenians, are herbivores, feeding on aquatic plants with their mobile lips and crushing teeth. Other marine mammals, such as the pinnipeds, display a variety of behaviors associated with mating, ranging from bloody dominance battles among males that compete for priority access to females to species stationed underwater engaging in complex vocal displays to attract females swimming past. Reproduction in marine mammals also differs; most give birth to a single offspring annually but in some species, including sirenians and nearly all whales, reproductive cycles are

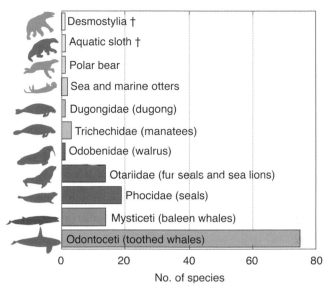

Figure 1.1. Diversity of marine mammals. Shading indicates major lineages.

separated by several years, an important factor to consider in their conservation and management strategies.

Many more marine mammal species existed in the past, some with no living counterparts. For example, extinct sloths and bizarre hippo-sized desmostylians, both herbivores, foraged in aquatic ecosystems. The number of species of marine mammals probably reached its maximum in the middle Miocene, 12–14 million years ago, and has been declining since then.

In this chapter, I present a brief introduction to the naming and classifying of marine mammals, the process of forming new marine mammal species, and factors responsible for their distribution. Chapter 2 provides a geologic context for interpreting the life and evolutionary times of marine mammals. In chapters 3–5, the evolutionary history, diversification, and adaptations of the major lineages of marine mammals are described. The final chapter, chapter 6, reviews the ecology and conservation of marine mammals.

DISCOVERING, NAMING, AND CLASSIFYING MARINE MAMMALS

The diversity of marine mammals makes their classification a challenge. The universal language of biology is **taxonomy,** which includes the identification, description, naming, and classification of organisms. Also, taxonomy plays an important role in conservation biology since before you can conserve organisms, you have to be able to identify what it is you intend to conserve. Although we often hear more about vanishing species, a number of new marine mammal species have also been discovered. For example, in the last decade two new species of baleen whales have been described: Omura's whale (*Balaenoptera omurai*) from the Indo-Pacific and a right whale (*Eubalaena japonica*) from the North Pacific. Among toothed whales, several new species of beaked whales (*Mesoplodon perrini* and *Mesoplodon peruvianus*), the Australian snub-fin dolphin (*Orcaella heinsohni*), and the narrow-ridged finless porpoise (*Neophocaena asiaorientalis*) have been described.

Common and Scientific Names

Marine mammals are given names and classified in much the same way as all organisms are named and classified. One problem in taxonomy is that the same common name is often applied to different animals. For example, the name "seal" has been applied to both sea lions and fur seals (or otariids) and seals (or phocids), which are two very different pinniped lineages. Another problem is that different common names can be applied to the same species. For example, the names "harbor porpoise" and "common porpoise" have been both applied to *Phocoena phocoena*. For these reasons, and since all species have a single, unique **scientific name**, it is more important to remember the scientific name than the common name. The scientific name of a species consists of the genus name and the species name and follows a set of rules of nomenclature developed by Carl von Linne, better known as Linnaeus, in the mid-1700s. In the previous example, following the Linnaean system of nomenclature, the harbor porpoise has two names: the first indicating that it belongs to the genus *Phocoena* (Latin for "pig fish") and the second, specific name, *phocoena*. Note that the first name is capitalized but that the second name is not.

DNA Bar Coding: Species Discovery and Conservation

Species-level differences between organisms encode genetic information (that is, changes in DNA). In much the same way as barcodes are used to uniquely identify commercial products in everyday life, **DNA bar coding** makes use of DNA sequences as unique identifiers of species (fig. 1.2). Given a reference database of sequences from validated specimens (identified by experts from diagnostic skeletal material or photographs), unknown specimens can be identified as belonging to a particular species. Application of DNA bar coding to the taxonomy of a poorly known family of beaked whales (Ziphiidae) resulted in the correct identification of previously misidentified specimens.

DNA bar coding also has important uses in conservation for the genetic identification of illegally imported animal or plant products. For example, DNA analysis of whale products (for example, meat and oil) found in retail market places in Japan, Korea, and the United States revealed the illegal trade of protected endangered species.

RECONSTRUCTING THE HIERARCHY OF MARINE MAMMALS

The Linnaean system organizes groups of organisms (for example, species) into higher categories or ranks (that is, families, orders, classes, etc.). The species is the basic, smallest level of biological classification. For example, the species *Phoca vitulina* is grouped into a larger unit of related species, the genus *Phoca*, which is in turn grouped into even larger hierarchies, such as Phocidae (seals) and Pinnipedia (including Otariidae, Odobenidae, and Phocidae). Given the arbitrariness of all ranks above the species, however,

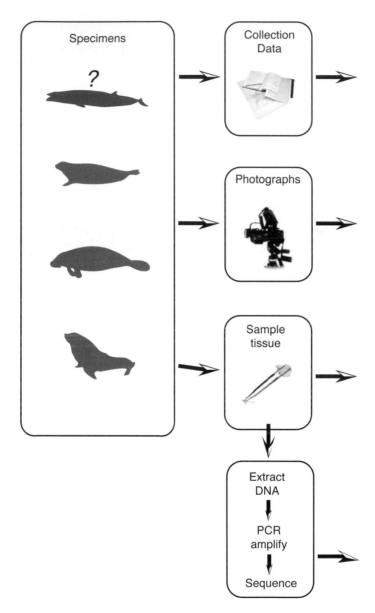

Figure 1.2. Steps involved in DNA barcoding: specimens, laboratory analysis, and database.

Web-accessible Data
and DNA Barcode

Sample # _____   Date _____

Location _____   Collector _____

Confirm identification
as *Megaptera novaeangliae*

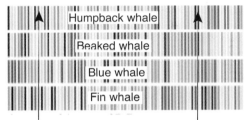

Humpback whale

Beaked whale

Blue whale

Fin whale

Compare to DNA Library

 ATAGGTGCAGAGACTCGACGGAAGCTATTCTAACGAATGAATATCATTT

some biologists have offered compelling arguments for the elimination of ranks above the species level altogether. However, regardless of whether ranks are employed, organisms can be organized into nested hierarchies based on the distribution of their shared features or characters. The reason for this underlying pattern of nested hierarchy was recognized by Charles Darwin in his 1859 masterpiece *The Origin of Species*, and attributed to common descent with modification—that is, **evolution**. The hierarchical nature of life reflects the tree-like nature of the history of life.

Characters are diverse, heritable attributes of organisms that include DNA sequences, anatomical features, and behavioral traits. Any characters that are shared by two or more species that have been inherited from a common ancestor are said to be **homologous**. For example, think of a bird wing and a seal flipper. They display similarities and differences. Although the forelimbs of a bird and a seal have different functions—one is employed in flying and the other is used for swimming—it is their similarities (that is, basic limb structure and bone relationships) that we are most interested in. We refer to this as a homologous similarity. Because homologous characters show evidence of inheritance, they are useful to determine evolutionary relationships among organisms. In this case, a bird wing and seal flipper are similar because they inherited this similarity from a common tetrapod ancestor. Homologous characters are also known as **synapomorphies**. Synapomorphies are **derived** characters shared among organisms. A derived character is one that is different from the ancestral character. For example, all tetrapods share four limbs; however, pinnipeds, a more inclusive group of tetrapods, share a more recent common ancestry and they can be distinguished from other tetrapods by possession of the derived character of limbs modified into flippers. Not all characters are evidence of relatedness. Similar traits in organisms can develop for other reasons, such as ecology. For example, the flipper of a seal and the flipper of a whale are not homologous because they evolved independently from the forelimbs of different ancestors—that is, the flipper of a sea lion is derived from carnivorans (for example, otters, bears, and weasels) whereas the flipper of a whale evolved from artiodactyls

TABLE I.I

Summary of the distribution of a few pinniped characters.

Taxon	Lacrimal Absent	Flippers	Orbital Maxilla	Reduced Claws	Tusks
Arctoids (outgroup)	0	0	0	0	0
Enaliarctos	?	√	?	0	0
Seal	√	√	√	0	0
Walrus	√	√	√	√	√
Sea lion	√	√	√	√	0

(even-toed ungulates like cows, pigs, and hippopotamuses). This is known as an **analogous** similarity; two characters are analogous if they have separate evolutionary origins. This is known as **convergent evolution**.

Derived characters are distributed hierarchically among a select group of organisms. Consider the example of flippers possessed by pinnipeds. Since all pinnipeds have both foreflippers and hind flippers, it follows that if one wanted to tell a pinniped from a nonpinniped (any other animal), one would need only observe that the pinniped is the one with four flippers. On the other hand, the character possession of foreflippers and hind flippers is not useful for distinguishing a seal from a sea lion—both have four flippers. To distinguish a seal from a sea lion, characters other than the presence of flippers must be used to identify subsets within the group that includes all pinnipeds.

We commonly use a branching diagram known as a **cladogram** or **phylogenetic tree** to visualize the hierarchies of derived characters within a group of organisms. The lines of a tree of life are known as **lineages** and represent the sequence of descent from parents to offspring over many generations. To illustrate how a tree is constructed, let's consider four pinnipeds: seal (phocid), walrus (odobenid), sea lion (otariid), and the fossil (*Enaliarctos*). For simplicity, I have selected traits that are either present (√) or absent (0) (table 1.1, fig. 1.3).

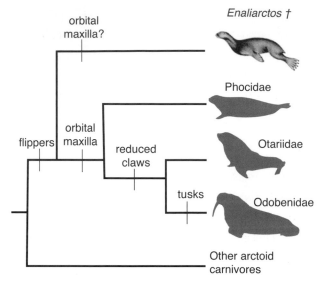

Figure 1.3. Distribution of character states among pinnipeds (restoration of stem pinniped by Mary Parrish).

A group of terrestrial carnivores, the arctoid carnivores (bears, weasels, and raccoons and their kin), are thought to have separated from the lineage leading to pinnipeds before the evolution of flippers. Therefore, arctoids are chosen as the **outgroup**—that is, outside the group of interest—for our analysis. As we will see in chapter 3, the extinct pinniped *Enaliarctos* is thought to have separated from the lineage leading to all other pinnipeds. Extant pinnipeds (and possibly *Enaliarctos*) differ from terrestrial arctoids in having the maxilla (upper jaw bone) form part of the lateral (side) and anterior (front) walls of the eye orbit. Walruses and otariids share a derived trait: the presence of reduced claws. We infer that reduced claws evolved in the common ancestor of walruses and otariids after that lineage separated from phocids. Walruses have one unique character in our list: the presence of tusks.

Any group of species that consists of all the descendants of a common ancestor is called a **monophyletic** group or a **clade**. In this example, walruses, phocids, and otariids are separate monophyletic clades that are united in a larger, more inclusive monophyletic, Pinnipedia. Two species or taxa that are each other's closest relatives are called **sister species** or sister clades. In this example, walruses and otariids are sister clades.

A group of species that does not include the common ancestor or all the descendants of a common ancestor is called a **nonmonophyletic** group. An example of a nonmonophyletic group is that of **river dolphins.** They differ from oceanic dolphins in inhabiting freshwater rivers and estuaries. Recent molecular data supports river dolphins as a nonmonophyletic group. Ganges river dolphins do not share the same common ancestor as other river dolphins (see also chapter 4). Most taxonomists agree that it is not appropriate to recognize nonmonophyletic groups as taxonomic units because they misrepresent evolutionary history.

Important concepts when defining members of a clade are stem and crown groups. A **crown group** is the smallest monophyletic group, or clade, to contain the last common ancestor of all extant members, and all of that ancestor's descendants. Extinct organisms can still be part of a crown group: for instance, the extinct northern fur seal (*Callorhinus gilmorei*) is still descended from the last common ancestor of all living otariids, so it falls within the otariid crown group. Some organisms fall close to but outside a particular crown group. A good example is *Enaliarctos*, which, although clearly pinniped-like, is not descended from the last common ancestor of all living pinnipeds. Such organisms can be classified within the **stem group** of a clade. In fig. 1.3, *Enaliarctos* is a stem group pinniped. All organisms more closely related to crown group pinnipeds than to any other living group are referable to the stem group. As living pinnipeds are by definition in the crown group, it follows that all members of the stem group of a clade are extinct; thus, stem groups only have fossil members.

ADAPTATIONS AND EXAPTATIONS

Adaptations are features that are common in a population because they provide improved function. For example, the ability of toothed whales to hear high-frequency sounds or echolocate is an adaptation for navigation and foraging. **Exaptations** are features that provide a function that is different from its original function. For example, it is hypothesized that the lower jaw of toothed whales may have arisen originally to transmit low-frequency sounds (as in some other mammals such as the mole rat, which hears ground vibrations) and later became specialized for transmitting high-frequency sounds. In this way, the lower jaw of toothed whales may be viewed as an exaptation for hearing high-frequency sounds, having initially functioned in low-frequency hearing.

WHAT IS A SPECIES AND HOW DO NEW SPECIES FORM?

One common but sometimes difficult question is how best to decide which particular species an organism belongs to. Another challenge is deciding when to recognize a new species. This is a question for the biologist, who discovers organisms that appear to be different from those that belong to already described species. Thus there are disagreements regarding what constitutes a species (that is, species concepts) as well as what are the best criteria for identifying species. Since species are often granted a greater degree of protection than populations, failure to recognize species may lead to inaccurate assessments of biodiversity.

For example, there is current debate over the species status of the killer whale. Traditionally, a single species of killer whale, *Orcinus orca*, found in all the world's oceans, has been recognized. There is now good evidence that several different species of killer whales exist in the northeast Pacific and Antarctic, based on differences in coloration, prey selection, habitat, and genetic data (fig. 1.4). Establishing appropriate taxonomic designations for killer whales is critical for understanding the ecologic impacts and conservation needs of these top marine predators.

Speciation is the process by which new species form from a common ancestor. In fig. 1.3, the branching of the tree denotes speciation

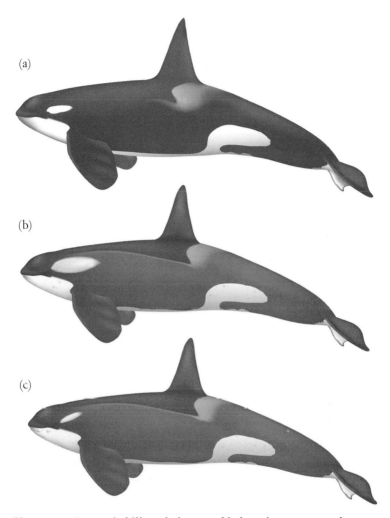

(a)

(b)

(c)

Figure 1.4. Antarctic killer whales, a and b, have been proposed as new species, with c proposed as a new subspecies (courtesy U. Gorter).

among various lineages of pinnipeds. There are three primary ways that new species form: (1) allopatric, (2) parapatric, and (3) sympatric speciation. In the most common type of speciation, **allopatric speciation**, new species arise from geographically isolated populations (fig. 1.5). In this type of speciation, a physical barrier prevents two or more groups

EVOLUTIONARY HYPOTHESES AND THE USES OF PHYLOGENIES

An important aspect of science is providing testable hypotheses to explain a set of data. Cladograms, or phylogenetic trees, are hypotheses of evolutionary relationship among a group of organisms. The phylogeny of extant pinnipeds is based on a small sample of characters. Typically, biologists construct phylogenetic trees using hundreds or thousands of characters. Large data sets require the use of computer programs to sort through millions or even billions of trees, searching for the best tree. One method of distinguishing among different hypotheses of relationship uses the principle of **parsimony,** which states that the preferred explanation of the observed data is the simplest explanation—that is, one that requires the fewest additional ad hoc assumptions.

Once a phylogenetic tree is reconstructed, it can be used to address wider evolutionary, ecological, and behavioral questions. For example, consider the evolution of locomotion in whales. If we map the various modes of locomotion onto whale phylogeny, we can hypothesize that the tail-based propulsion of extant whales in water evolved from initial use of fore and hindlimbs on land. This was followed by a pelvic phase that involved paddling with their feet (for example, *Ambulocetus*), a caudal undulation phase in which the tail and back were used (for example, *Kutchicetus*), and the final adoption of tail-based propulsion (dorudontines and crown cetaceans).

Phylogenies can help us determine conservation priorities. For example, the Ganges river dolphin lineage, formerly a diverse clade, is made up of only one extant member (*Platanista gangetica*). Among toothed whales, this species is an early diverging lineage and preserves ancestral character states of toothed whales, such as their origin in marine waters prior to invading present-day freshwater habitats. For this reason, on the basis of their evolutionary distinctiveness as well as other factors, including human activities, this lineage of river dolphins is critically endangered and is a high priority for conservation.

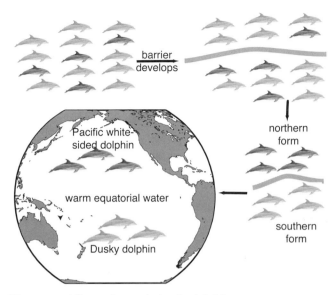

Figure 1.5. Allopatric speciation in dolphins.

from mating with each other regularly, so the lineage divides over time. In the case of marine mammals, isolation might occur because a barrier, such as warm equatorial water, divided a broadly distributed ancestral population inhabiting cool temperate water. An allopatric origin has been suggested for Pacific white-sided dolphins (*Lagenorynchus obliquidens*), which inhabit the Northern Hemisphere, and their sister species, the dusky dolphin (*L. obscurus*), which lives in the Southern Hemisphere. The two species are separated by warm equatorial water.

A special version of allopatric speciation is **peripatric speciation**. It occurs when a small population becomes isolated at the edge of a larger, ancestral population (fig. 1.6). The small population is referred to as the **founder population**. Elephant seals are an example of peripatric speciation. During the late 1800s, entire herds of northern elephant seals (*Mirounga angustirostris*) in California were slaughtered for the high oil content of their blubber. The Mexican government protected them on the Isla Guadalupe off the coast of Mexico. This small, isolated founder

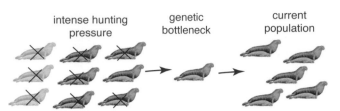

intense hunting pressure genetic bottleneck current population

Figure 1.6. Peripatric speciation in elephant seals.

population grew and eventually recolonized the mainland of California and Mexico. Although these magnificent animals have made an amazing comeback, the severe reduction in their population (termed a **genetic bottleneck**) in the late 1800s and early 1900s is still an important issue. Since all of the current northern elephant seals are descendants of the same 20–100 seals of the founder population, they are genetically very similar. Unfortunately, this lack of genetic variation makes them very vulnerable to new diseases or environmental changes.

In **parapatric speciation,** a new species arises within a continuously distributed population (fig. 1.7). There is no specific physical barrier to gene flow. The population is continuous but, nonetheless, the population does not mate randomly. Individuals are more likely to mate with their geographic neighbors than with individuals in a different part of the population's range. A possible example of parapatric speciation is provided by coastal and offshore populations of bottlenose dolphins (*Tursiops truncatus*). These bottlenose dolphin forms are morphologically and, in some cases, genetically distinct. Their habitats differ in various ways, including available prey, with the offshore form feeding on pelagic fish while the nearshore form eats shallow-water fish.

In the third major type of speciation, **sympatric speciation,** a new species arises within the range of the ancestral population (fig. 1.7). Like parapatric speciation, sympatric speciation does not require a geographic barrier to reduce gene flow between populations. Instead, some members of a population exploit a different niche, such as when feeding on a new prey item, which can promote reproductive

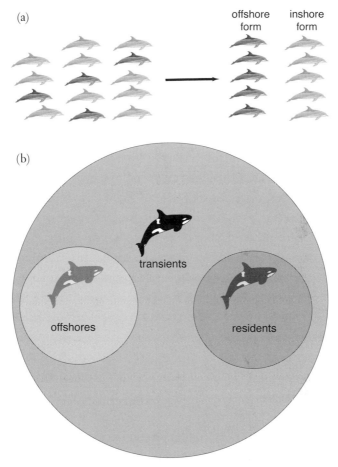

Figure 1.7. (a) Parapatric speciation in inshore and offshore bottlenose dolphins; (b) sympatric speciation in transient, resident, and offshore killer whales.

isolation among populations. **Resident, transient,** and **offshore killer whales** in the North Pacific provide an example of sympatric speciation. Resident populations occur in certain coastal regions and generally consume fish. Offshore populations inhabit waters farther from the coast and also feed on fish. Transient populations have the largest geographic range, overlapping with the other two types.

They feed exclusively on other mammals, such as dolphins and seals. Recent genetic data supports species status for transients and sub-species designations for resident and offshore populations of killer whales.

A fourth mode of speciation, **hybridization**, the successful mating between individuals of two different species, is relatively rare in animals but it is observed frequently in plants and is the dominant type of speciation in many agricultural plants (that is, corn, wheat, oat). In cases such as this, hybridization has been an important source of evolutionary novelty. In one of the few examples of hybrid speciation in animals, DNA evidence has revealed that polar bears arose by recent hybridization with an extinct population of brown bears. This hybridization event occurred not in Alaska, as previously thought, but in the vicinity of present-day Britain and Ireland during the last ice age 20,000–50,000 years ago. Hybrids in both captivity and the wild have been reported in nearly one-half of known marine mammal species, with the majority described among otariid pinnipeds, especially southern fur seals (*Arctophoca* and *Arctocephalus* species). Reasons for the high rate of hybridization have been attributed to bottleneck populations resulting from near decimation of fur seal populations during Antarctic sealing in the 1800s. In these cases, hybridization would have been favored since it provides an opportunity for gene flow between otherwise isolated gene pools (for example, Antarctic and Subantarctic fur seals (*Arctocephalus gazella* and *Arctocephalus tropicalis*). An increase in hybridization in Arctic species, such as the polar bear, has been attributed to melting sea ice and interbreeding, which results from isolated populations coming into contact. Hybridization, in this case driven by human activities, has a negative effect—a potential to reduce genetic diversity—since later generations may be less fit than their ancestors and interbreeding could mean the extinction of rare populations or species.

WHERE DO THEY LIVE AND
WHY ARE THEY WHERE THEY ARE?

Consideration of why mammals returned to the sea necessitates review of the physical and ecologic factors of the marine environment that influence life in the sea today. These physical factors include ocean temperature, depth, salinity, and circulation patterns and ecological requirements of species, such as food availability and abundance.

Sea surface water temperature patterns vary geographically and seasonally and affect the distribution of marine mammals. Surface ocean temperatures tend to be highest at the equator and decrease toward the poles. This poleward gradient of surface ocean temperatures establishes marine climate zones. Sea ice forms only in polar and subpolar zones. Seasonal cycles of freezing and melting of sea ice limit access to high latitudes from most marine mammal species to only the warmest summer months.

Marine Biodiversity Hotspots

Marine mammals concentrate in **marine biodiversity hotspots** much like those that exist on land. The specific location of these hotspots, for example, along continental shelves, sea mounts, and coral reefs in areas of increased food availability has conservation implications since they can provide the basis for establishing open ocean marine reserves.

Modeling approaches are used to generate predictions of global distributions of marine mammals. These estimates use the environmental tolerance of a species with respect to selected factors such as depth, salinity, temperature, primary productivity, and its association with sea ice or coastal areas. An international team of ecologists using a model of species distributions and oceanographic data revealed that current hotspots of marine mammal diversity are concentrated in the temperate latitudes of both hemispheres (for example, the Pacific coasts of North America, the waters around New Zealand, and the Galapagos Islands).

The availability and abundance of food for marine mammals is established by a number of factors, including the number of **trophic levels** between the primary producer and the marine mammal consumer (see also chapter 6) and by rates of **primary production**. Sirenians are the only marine mammals to feed directly on primary producers (sea grasses and algae), whereas some pinnipeds and whales feed on prey five or more trophic levels removed from primary producers. Rates of primary production in the ocean can vary over geographic areas and also between seasons. Seasonal variation in primary production is related to differences in light intensity, water temperature, nutrient abundance, and grazing pressure. A dramatic increase in primary production, especially diatoms, begins in the spring and continues through the summer in subpolar and polar seas (fig. 1.8).

SPECIATION AND TIMING: MOLECULAR CLOCK

Sometimes biologists want to understand not only the order in which evolutionary lineages split, but also the timing of those events. Time estimates come from the concept of a **molecular clock**. The number of changes that accumulate in gene sequences between any pair of species is proportional to the time since they last shared a common ancestor. A molecular clock must be set or calibrated using independent data such as the fossil record, which provides times of lineage divergence or biogeographic data (such as dates for separation of ocean basins in the case of marine mammals).

Attempts to date the molecular divergences of various marine mammal lineages in most cases agree with the fossil record; however, there are a few divergence dates that have proved controversial. For example, the divergence dates for hippos and whales are considerably younger than the first appearances of whales. This discrepancy of more than 40 million years may be due to a number of different reasons, including inadequate sampling of both genes and fossils and is the subject of additional work.

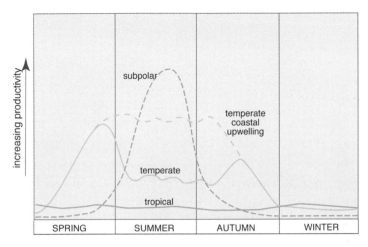

Figure 1.8. General patterns of seasonal variation in marine productivity for four different marine production systems (from Berta et al. 2006).

Breeding Areas and Migratory Corridors

Although a few pinnipeds, such as the Weddell and crabeater seals, exploit high-latitude areas year-round, mysticetes or baleen whales typically undertake intensive summer feeding in subpolar and polar seas, followed by long-distance migrations to low-latitude calving grounds in winter months. Eastern Pacific gray whales arguably accomplish the longest annual migration of any mammal, covering 15,000–20,000 km (9,000–12,000 mi) in their migration from feeding grounds in the Bering, Chukchi, and Okhotsk seas to warmer, sheltered breeding and calving grounds along the coast of Mexico (fig. 1.9). In the past, however, gray whales did not migrate, since their major feeding grounds disappeared during the last glacial period. During that time sea level is estimated to have dropped by nearly 120 m (400 ft), which eliminated 60 percent of the Bering Sea platform (see also chapter 6).

Humpback whales are grouped into different populations that live in three general areas: the North Pacific, the Atlantic, and the Southern Ocean. Because seasons are reversed on either side of the equator, Northern and

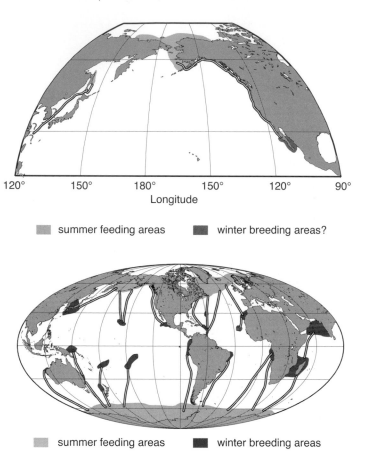

Figure 1.9. Distribution of gray whale (top) and humpback whale (bottom) feeding and breeding areas and the migration routes linking them (from Berta et al. 2006).

Southern Hemisphere populations of humpback whales probably never meet; those in the north travel toward their breeding grounds in tropical waters as those in the south are travelling toward the pole to feed, and vice versa (fig. 1.10). For example, North Pacific humpback whales migrate from Alaska to Hawaiian breeding grounds at the same time that Southern Ocean humpback whales are travelling to Antarctic feeding grounds.

In migratory species such as gray and humpback whales, the newly

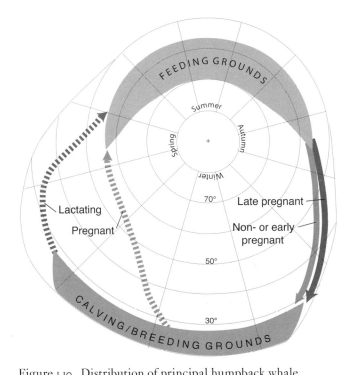

Figure 1.10. Distribution of principal humpback whale
feeding and breeding areas and migration routes linking
them (from Berta et al. 2006).

pregnant females are the first to leave the breeding ground and return to
summer feeding areas. A pregnant female must rapidly put on enough
fat to sustain her and her growing fetus through the coming year. Late
pregnant cows are among the first to return to the breeding and calving
grounds, since early calving allows more time for a calf to grow before
it too must migrate.

WHY DO SOME WHALES AND SEALS MIGRATE?

The function of the annual migrations of mysticetes is unknown but, given
that it is such a huge energy commitment, it likely involves several factors, such as reduced risk of killer whale predation on vulnerable calves,

thermoregulatory benefits to calves born in warmwater, and insufficient food in the feeding areas during the winter. Whale migrations, however, are not necessarily a well-defined procession of animals moving north or southward during a specific time of year. For example, Bryde's whales, north Atlantic minkes, and pygmy right whales live in temperate waters in all seasons, presumably because they can find enough food year-round to sustain them. Bowhead whale migrations depend on the condition of Arctic pack ice, which varies from year to year. Most toothed whales do not undergo long distance migrations. Sperm whales, however, are an exception and adult males leave their family groups in equatorial waters and travel to feed in polar waters in the summer (see also chapter 4).

In addition to the annual migrations of most mysticetes, several pinnipeds, including harp and hooded seals, northern and southern elephant seals, and possibly Weddell seals, also undertake seasonal migrations. In the case of elephant seals, two round-trip migrations are made yearly between nearshore island breeding rookeries and offshore feeding areas (see also chapter 3).

The distribution of food resources is also affected by seasonal shifts in water circulation and temperature. The ocean's layers of water have different temperatures. In areas of **upwelling**, wind-driven, dense, cool, nutrient-rich water replaces warmer water at the sea surface. Regions of upwelling are most commonly located along western edges of continents, areas occupied by numerous marine mammal species. Also present in coastal waters are **deep scattering layers** made up of vertically migrating zooplankton such as krill. These animals avoid daylight to escape predation by visual hunters and come up to feed at night, thus creating an increased density of food in surface waters that can be efficiently captured by marine mammals.

El Niño (El Niño Southern Oscillation or ENSO) events are disruptions in ocean and atmospheric circulation occurring at irregular intervals (typically every two to seven years) that result in the warming of surface waters in the eastern tropical Pacific, which blocks the transport of deeper, nutrient-rich water from below (fig. 1.11). El Niño events

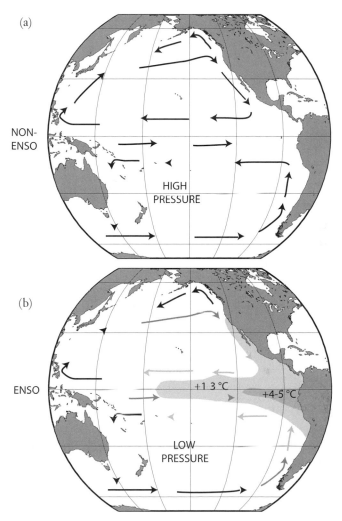

Figure 1.11. Generalized Pacific Ocean surface currents during non-ENSO (a) and ENSO (b) conditions (modified from Berta et al. 2006).

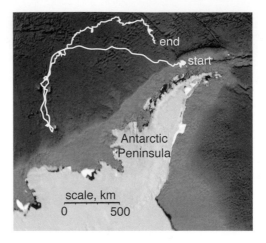

Figure 1.12. Satellite tracks
for a southern elephant seal
(from Costa et al. 2010).

have catastrophic effects on marine mammals, especially those living in
tropical and subtropical areas. For example, during the 1982–83 ENSO
event, 100 percent mortality of Galapagos fur seals and sea lions was
reported.

HOW DO WE KNOW WHERE THEY GO?

Study of the diving and feeding behavior of marine mammals has been
greatly enhanced by continuing advances in **satellite telemetry** and
microprocessors, which provide information on where animals go and
what they do underwater. Data are collected from sensors that measure
various parameters such as water temperature, depth, light intensity,
swimming pattern and velocity, heart rate, and body temperature. These
sensors transmit data to orbiting communication satellites that relay the
information to receiving stations on Earth, enabling the animals to be
tracked wherever they go. Think of these instruments as flight recorders for marine mammals (fig. 1.12). Tracking projects in progress, such
as the Tagging of Pacific Predators (TOPP), which began more than
a decade ago, have provided much data on the underwater activities
of various marine animals, including elephant seals. In another project, scientists from NOAA's Southwest Fisheries Science Center placed

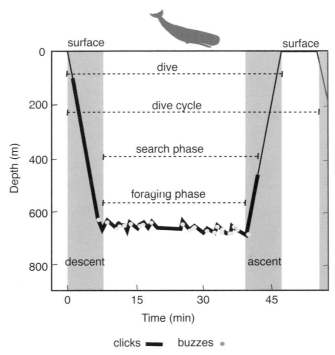

Figure 1.13. DTAG data on the diving and feeding behavior of a sperm whale (modified from Watwood et al. 2006).

satellite tags on fur seals, Weddell seals, and penguins to track their movements in Antarctica over the austral winter (our summer), when weather conditions make it nearly impossible for us to monitor them. Pinnipeds have proven particularly amenable to study since they **haul out** to breed and molt at predictable times and locations, making it feasible to attach and to retrieve animal-borne instruments.

One type of tag that has yielded new insights on the behavior of marine mammals is the **digital acoustic tag** (or DTAG). DTAGS are attached to the animal by a suction cup and record the 3D environment, including sound and motion (orientation), of the animal (fig. 1.13). DTAGs can also monitor marine mammal vocalizations, such as those used when diving and foraging, as well as their responses to human activity. For the entirely aquatic cetaceans and sirenians, the logistics are more difficult and researchers must follow tagged animals at sea to recover the released tag.

The use of photographic techniques to identify individual marine mammals is based on patterns of scars, natural pigmentation, callosities, or barnacle patches on the skin of a marine mammal. The technique, termed **photoidentification** or "photo ID," involves photographing a visually unique part of the animal, such as its head, dorsal fin, or fluke, and then collecting and cataloging these images so they can be compared to photographs of the same animal taken at another time or place. Newer and faster methods for digitizing images and computer-based retrieval and matching systems have been developed in recent years.

CHAPTER TWO

Past Diversity in Time and Space, Paleoclimates, and Paleoecology

In this chapter, I begin by introducing fossils and a geologic time frame and providing a context for interpreting marine mammal fossils and possible causes of their origin and diversification. Next, I consider the evolution of marine mammal communities through space and time and examine what may have led mammals back to the sea. Finally, I consider the use of a new technology, stable isotopes, for ecologic studies ranging from reconstructing paleotemperatures and climate change to documenting diet and foraging ecology among both extant and extinct marine mammals.

FOSSILS AND TAPHONOMY

In order to know where and when marine mammals originated and diversified, we have to study their remains or **fossils**, usually best preserved as hard parts such as bones and teeth. Fossils of marine mammals are known from as far back as 55 million years ago (Ma), and study of them has revealed much about their past life, evolution, and ecology. Just as important as the remains of once living animals is the context or environmental setting when the organism died and was buried. **Taphonomy** is the study of those changes that happen to organisms after death. In the case of marine mammals, such changes have revealed whether bones were carried off by other animals, with skeletal elements widely dispersed, or whether the animal was travelling in a herd that suffered a catastrophic mass death of many individuals in a short time, such as one caused by a dramatic drop in sea level, in which the bones of different individuals are found together, often relatively undisturbed. Sediments enclosing the fossil bones can also reveal the environment of deposition—for example, whether the marine mammal lived on land or in marine, brackish, or fresh water.

So where do you find fossils of marine mammals? Fossil marine mammal localities are known worldwide from the tropics to the poles and on every major continent, with most in the northern temperate regions, as discussed later in this chapter. An example of a spectacular marine mammal fossil locality is the Wadi Al-Hitan or Valley of Whales in Egypt (fig. 2.1). Known since the early 1900s, this is a **UNESCO World Natural Heritage** site today where more than 400 whale and sea cow skeletons, as well as sharks, bony fish, crocodiles, and sea turtles have been found. The site spans more than 10 km (6.2 mi) of sediments deposited during the middle Eocene between 40 and 39 Ma. The presence of many juvenile skeletons suggests that the area was a shallow bay, perhaps a calving ground for various marine mammal species.

THE DISCOVERY OF THE FIRST FOSSIL MARINE MAMMAL (A WHALE)

Although a portion of a whale jaw from Malta was illustrated in the seventeenth century, the widely cited "earliest" record of marine mammals are whale bones from Louisiana published in 1834 by Richard Harlan, the American geologist and paleontologist. He described large vertebra that had been washed out of a river bank. However, he did not recognize his find to be from a whale. Rather, he identified it as a dinosaur and proposed to call the animal "the king of lizards" or *Basilosaurus*. Among additional fossil remains of similar animals found in Alabama were teeth that were examined by a well-known anatomist, Richard Owen. In 1839, Owen proposed a new name for the bones, which he recognized as deriving from an

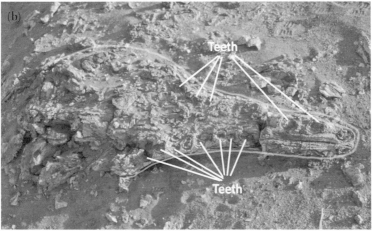

Figure 2.1. Wadi Al-Hitan (Valley of Whales), Egypt: (a) a whale skeleton weathering at surface and (b) a close-up of *Basilosaurus* skull showing palate (courtesy Mark Uhen).

Figure 2.2. First marine mammal fossil discovered (from Heuvel-mans, 1968).

extinct whale, *Zeuglodon cetoides.* However, the first published name takes precedence, in this case, rendering these specimens *Basilosaurus.*

A skeleton of *Basilosaurus* was found in Alabama in 1842; it had a partial skull, forelimb, and vertebral column extending 65 ft. This discovery prompted the fossil collector Albert Koch to visit Alabama and collect the remains of at least five different individuals of two different genera of early whales. He strung the bones together and "reconstructed" a single specimen 114 ft long. He claimed that such a skeleton belonged to a "sea serpent." The "skeleton" was exhibited in New York and Europe. Later, more careful examination showed that the skeleton in question was a composite of at least five different specimens of the ancestral whale *Basilosaurus* (fig. 2.2)

THE IMPORTANCE OF FOSSILS

The results of changing climates, geography, and environments are preserved in the fossil record of marine mammals, providing a valuable perspective and, in many cases, lessons for the future. For example, a study of northern fur seal ecology revealed a major ecological shift in the past. Based on harvest data from archaeological sites, northern fur seals

(*Callorhinus ursinus*) had more temperate breeding colonies in the past (for example, ranging from California to the Aleutian Islands) compared to modern high-latitude colonies (that is, Pribilof and Commander Islands, Bering Sea). The extinction of these temperate-latitude colonies is attributed to both human hunting and climatic factors and it is unknown which played a more important role. A potential conservation strategy involves the reestablishment of temperate-latitude breeding colonies and ecosystems, which could help offset the global decline of the species.

HOW DO WE KNOW THE AGE OF A FOSSIL?

Fossils are found in rocks, and the age of rocks is measured in term of relative and absolute dating. Using **relative dating**, the sequence or order of events can be determined based on the occurrence of a fossil in a layer of rocks, and the relative age of the fossil can be determined to be older or younger than the rocks that contain it. **Absolute dating** provides determination of a specific date of a rock containing the fossil, but it can only be applied if rocks can be found that contain certain minerals bearing radioactive forms of elements. These radioactive isotopes break down or decay according to a predictable timetable. If the decay is recorded in the mineral or rock, we can infer the age of the rock and therefore the fossil.

HOW DO WE KNOW WHERE
MARINE MAMMALS WERE?

The shape and position of continents have changed through time. These movements of the continents, called **continental drift**, have had a major effect on the evolution of marine mammals. We know through the theory of **plate tectonics** that the earth's surface is divided into a series of large plates that move relative to one another. As a result, oceans open and close and continents move, sometimes separating, colliding, or sliding past one another.

Changes in the positions of continents have changed patterns of

GEOLOGIC TIME SCALE

In order to understand global events, geologists must talk the same language when they date rocks and the fossils contained within them. This has resulted in development of a standard relative **geologic time scale** that is divided into a hierarchy of time intervals, ranging from oldest to young-

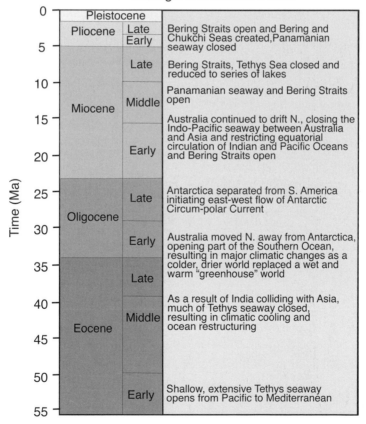

Figure 2.3. Geologic time scale with major geologic, climatic, and marine mammal events (modified from Berta et al. 2006 and Fordyce 2008).

est, that correspond to important changes in the history of life on Earth. Fig. 2.3 depicts a geologic time scale that lists major geological and climatic events in the evolutionary history of marine mammals.

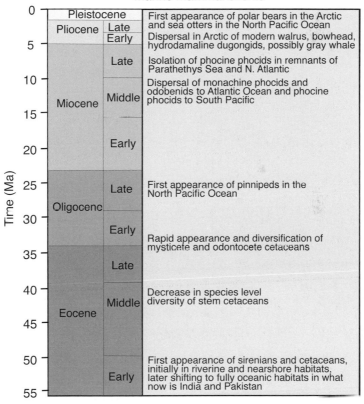

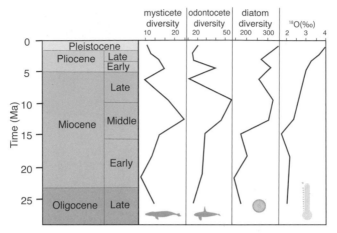

Figure 2.4. Cetacean and diatom diversity and ¹⁸O levels through time (modified from Marx and Uhen 2010).

ocean circulation and, consequently, food availability. For example, the radiation of baleen and toothed whales that began in the late Oligocene (38 Ma) has been associated with the breakup of southern continents and restructuring of ocean circulation patterns, which increased primary productivity and upwelling of nutrient-rich water. Important indicators of productivity are single-celled **diatoms**, a major group of phytoplankton (fig. 2.4). The dramatic rise of crown mysticetes and odontocetes (Neoceti) during the later Tertiary (15–2.5 Ma) has also been linked to primary productivity.

MARINE MAMMAL DIVERSITY AND COMMUNITIES THROUGH TIME

Diversity estimates of marine mammals through time show that the greatest generic diversity of cetaceans, pinnipeds, and sirenians occurred during the middle Miocene (12–14 Ma). It would be best to tabulate the number of fossil marine mammal species over time rather than the number of genera. However, the number of species, in many cases, is based on incomplete material that is often not diagnostic at the species level. It is likely that generic diversity

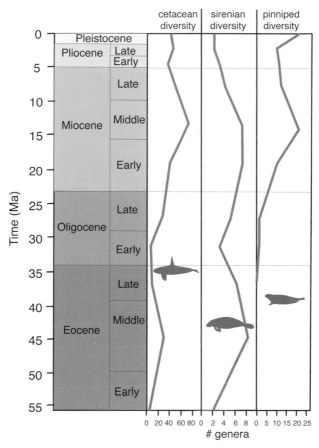

Figure 2.5. Generic diversity of marine mammals through time (adapted from Uhen 2007).

is a conservative estimate of actual species diversity. For whales and sirenians, generic diversity was double what it is today, while for pinnipeds, a comparable present-day diversity existed during the middle Miocene (fig. 2.5).

Reconstructed continents and ocean basins for each major time period with major marine mammal localities plotted are shown in figs 2.6, 2.8, 2.10, and 2.12. Communities of various marine mammals during each of these major time intervals (although not necessarily found in the same geographic areas) are shown in figs. 2.7, 2.9, 2.11, and 2.13.

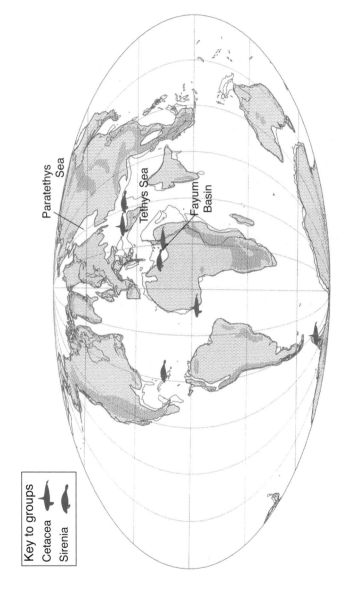

Figure 2.6. Major marine mammal fossil localities during the early–late Eocene (modified from Berta et al. 2006 and Fordyce 2008).

Figure 2.7. Representative marine mammal community during the early–late Eocene, including (clockwise from top left) whales *Kutchicetus, Dorudon,* and *Rodhocetus,* and sirenian *Protosiren* (painted by Carl Buell).

Early–Late Eocene (54–34 Million Years Ago)

During this time, an extensive shallow Tethys Sea (named for a sea goddess of Greek mythology) stretched from the Pacific to the present-day Mediterranean Sea. By the end of the Eocene, India had moved northward to collide with Asia, closing much of the Tethys. Remnants of the Tethys Sea, known as the **Paratethys Sea**, existed to the west through what is now southern Eurasia (fig. 2.6).

By the end of the Eocene (34 Ma) in the Southern Hemisphere, Australia had moved north, away from Antarctica, opening part of the Southern Ocean—a circumpolar seaway connecting the southern portions of the Pacific, Atlantic, and Indian Oceans. Steep temperature gradients developed between cold Subantarctic and temperate waters, leading to a cooling of the Southern Ocean and initiation of glaciation on Antarctica.

Whales and sirenians originated and diversified on eastern shores of the Tethys Sea (what is now India and Pakistan) and the western tropical Atlantic (Caribbean; fig. 2.7). Among the best-known localities is the middle and late Eocene Fayum Basin in Egypt, including the Valley of Whales discussed earlier, which has yielded a considerable diversity of fossil sirenians and whales. By the end of the Eocene, closure of the Tethys Sea and concomitant global cooling resulted in a decrease in species-level diversity.

Latest Oligocene–Early Miocene
(33–17 Million Years Ago)

Antarctica and South America separated in the late Oligocene (30–23 Ma), allowing west-to-east flow of the newly developed **Antarctic Circumpolar Current** (fig. 2.8). This current isolated Antarctica and probably allowed the Antarctic icecap to expand, global climates to cool, and oceans to become more mixed, resulting in their increased productivity. In addition, a major global sea level rise is likely to have expanded continental shelf habitats.

The **Central American** or **Panamanian Seaway** separating North and South America, open for most of the Cenozoic (until 11 Ma but open again from 6 to 4 Ma), allowed the exchange of waters and biota between the equatorial Pacific and Atlantic oceans.

Odontocete and mysticete whales or Neoceti originated in the Oligocene, and an explosive radiation of both lineages in the late Oligocene has been attributed to increased zooplankton productivity associated with restructuring of the Southern Ocean ecosystem. Among the best-known fossil localities is the Waitaki Valley, New Zealand (fig. 2.8).

The Central American seaway likely served as the dispersal route for monachines (that is, elephant and monk seals) in coastal Peru and diverse phocids and walruses along the eastern Atlantic.

The North Pacific was the ancestral home of a bizarre, wholly extinct group of semiaquatic herbivorous marine mammals related to sea cows,

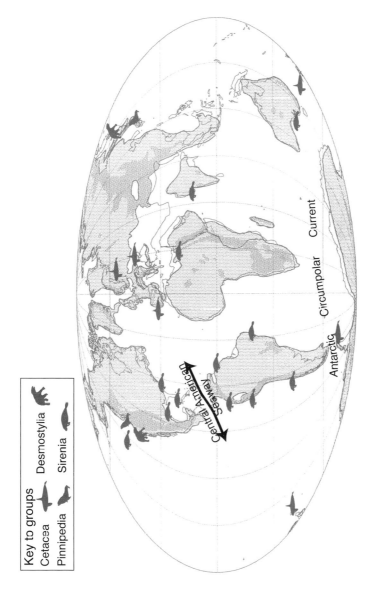

Figure 2.8. Major marine mammal fossil localities during the late Oligocene–early Miocene (modified from Berta et al. 2006 and Fordyce 2008).

Figure 2.9. Representative marine mammal community during the latest Oligocene–early Miocene, including (clockwise from top left) desmostylian *Paleoparadoxia*, odontocete *Simocyon*, and pinniped *Enaliarctos* (painted by Carl Buell).

the desmostylians. Cold-adapted desmostylians grazed in shallow tidal zones or offshore reefs and stem pinnipeds, such as *Enaliarctos*, apparently diversified in shallow bays or inland seas but remained restricted to the eastern North Pacific (fig. 2.9). A lineage of cold water-adapted dugongids, the hydrodamalines, occupied the Pacific coast during this time.

Middle Miocene–Pliocene (16–1.6 Million Years Ago)

The late Miocene (5–6 Ma) collision of Africa with the Iberian Peninsula in southwestern Europe marked the large-scale drying of the Mediterranean Sea known as the **Messinian Salinity Crisis** and isolation of the Paratethys as a brackish inland sea (fig. 2.10). The flow of warm water from the Mediterranean to the Atlantic was shut off, which intensified the cooling of the North Atlantic. Resulting steep thermal

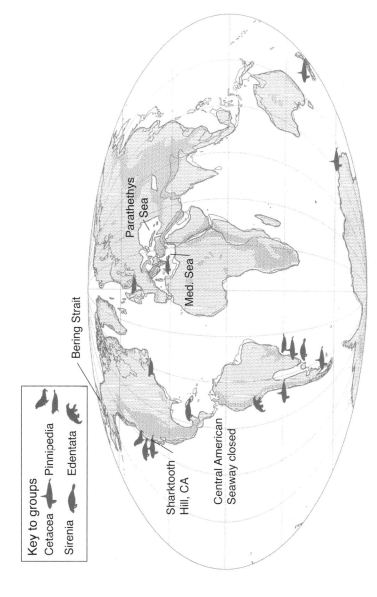

Figure 2.10. Major marine mammal fossil localities during the mid-Miocene–Pliocene (data from Fordyce 2008).

Key to groups
Cetacea   Pinnipedia
Sirenia   Edentata

Bering Strait

Parathethys Sea

Med. Sea

Sharktooth Hill, CA

Central American Seaway closed

gradients were likely associated with increased levels of upwelling and phytoplankton productivity in the North Atlantic, and a similar chain of events is postulated for the North Pacific. Also in the late Miocene, an ice cap formed in West Antarctica, resulting in a major cooling of the Southern Ocean.

In the Northern Hemisphere, the **Bering Strait**, a seaway between Alaska and Siberia, first opened as the result of plate tectonic activity during the latest Miocene to earliest Pliocene (5.5–4.8 Ma). The flow of surface marine water through the Bering Strait reversed during the mid-Pliocene (3.6 Ma) and the resulting south-to-north flow established the modern Arctic Ocean circulation pattern.

In the middle Pliocene (3–4 Ma), the Central American seaway closed, with emergence of the Isthmus of Panama, cutting the Caribbean-Pacific links. In South America, uplift of the Andes during the later Miocene and early Pliocene (4–6 Ma) created erosion and runoff of nutrients into river systems, which resulted in an abundance of abrasive grasses.

The earliest confirmed phocids are known from the western North Atlantic and had established a circum-Atlantic distribution by the late Miocene that included the Mediterranean Sea and Parathethys Seas. Monk seals are believed to have originated in the Mediterranean (*Monachus monachus*), with dispersal east in the Caribbean (*M. tropicalis*) and then west in the North Pacific (*M. schauinslandi*).

Various marine mammals, including both pinnipeds and whales, and their prey (that is, abalone, salmon, and rockfish) radiated in the cool waters of the North Pacific. Walruses may have evolved earlier but are clearly present in the middle Miocene in the North Pacific. The later opening of the Bering Strait was associated with trans-Arctic biotic interchange that likely included Atlantic and Arctic mollusks and their chief predators, odobenine walruses. Both toothed and baleen whales also underwent extensive diversification in the Southern Ocean, the result of an increase in diatom-based production fueled by upwelling and deep-water mixing induced by the Antarctic Circumpolar Current.

Figure 2.11. Representative marine mammal community during the mid-Miocene–Pliocene, including (clockwise from top left) pinniped *Allodesmus kernensis,* baleen whale *Balaenoptera* sp., desmostylian *Desmostylus hesperus,* walrus *Valenictus chulavistensis* (painted by Carl Buell).

In the South Pacific from the late Miocene to the late Pliocene, another marine mammal lineage, large aquatic ground sloths radiated, grazing on algae or sea grasses. Manatees evolved during the middle Miocene (14 Ma) in coastal rivers and estuaries in South America. By the late Miocene and early Pliocene, they had adapted to abrasive freshwater grasses of rivers, eventually gaining access to the Amazon Basin, where they speciated (see also chapter 5).

In addition to diversification of pinnipeds in the late Miocene, this was also the time of global diversification of toothed and baleen whales and sirenians. On the Pacific coast, the mid-Miocene Sharktooth Hill bonebed and Plio-Pleistocene San Diego Formation in Southern California have yielded an abundant assemblage of marine mammals including whales, pinnipeds, and sirenians (figs. 2.10 and 2.11).

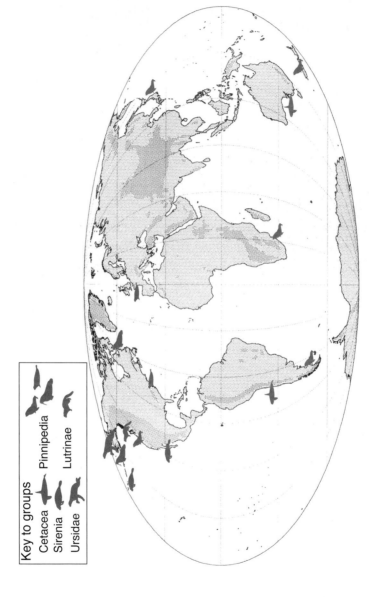

Figure 2.12. Major marine mammal fossil localities during the Pleistocene (data from Fordyce 2008).

Figure 2.13. Representative marine mammal community during the Pleistocene, including (clockwise from top left) sea otter (*Enhydra lutris*), Risso's dolphin (*Grampus griseus*), polar bear (*Ursus maritimus*), northern fur seal (*Callorhinus ursinus*), dugong (*Dugong dugon*), walrus (*Odobenus rosmarus*), and bowhead (*Balaena mysticetus*; painted by Carl Buell).

Pleistocene (1.6 Million–10,000 Years ago)

During this time interval, the warm and cold fluctuations in the climate begun during the Pliocene continued. Changing sea levels during glacial and interglacial intervals typified various regions during this period (fig. 2.12).

Marine mammal communities included living representatives of all lineages (fig. 2.13). Extant walruses (*Odobenus*) were present in both the North Atlantic and North Pacific and may have dispersed along either an Arctic route or via the Central American seaway. Phocine seals reached the North Pacific during the Pleistocene probably via an Arctic dispersal route. Likely glacial and interglacial oscillations resulted in reduced gene flow between isolated populations and promoted the formation of new species by allopatric speciation such as

CHANGING TEMPERATURES, CLIMATE, AND ECOLOGY

It is possible to determine how cold or hot the earth was in the past or the salinity of ancient oceans using **stable isotopes**. Stable isotopes, one of several forms of an element, are distinguished from others (that is, radioactive isotopes used in dating rocks) by being nonradioactive. Stable isotope values of oxygen, nitrogen, and hydrogen are also used to study diet and trophic level and habitat use. For example, many marine organisms (such as diatoms, discussed earlier) that take up silicate from ocean water to build their skeletons select one oxygen isotope in preference to another depending on water temperature. The ratio of oxygen 16 (^{16}O) to oxygen 18 (^{18}O) in fossil shells may indicate the water temperature at which the shells were formed. Thus in fig. 2.4 the oxygen isotope curve, a proxy for climate change, indicates variation in water temperature that parallels the diatom diversity curve. This is an indicator of productivity which suggests that cetacean diversity was driven by increased diatom-based primary production, which in turn effected climate (variation in ^{18}O values).

The isotopic concentrations in mineralized tissues of animals (for example, teeth, bone) reflect the isotopic composition of the food and water they ingest. Consequently, isotopic data can be used to study diet (carnivore vs. herbivore, freshwater vs. marine) and habitat preference (pelagic vs. benthic, nearshore vs. offshore vs. estuarine). For example, carbon and oxygen isotope data for sirenians reveal that extant manatees range from entirely freshwater to entirely marine, although an early fossil member of this lineage (that is, *Potamosiren*) likely foraged exclusively in freshwater.

the spotted seal (*Phoca largha*) the ribbon seal (*Histriophoca fasciata*) and the harp seal (*Pagophilus groenlandica*). In the Southern Hemisphere, monachines diversified in cool waters to produce the Antarctic lobodontine seal fauna, such as the leopard, Weddell, and crabeater seals of today. Otariid seals, notably the southern fur seals (*Arctophoca* and *Arctocephalus*), underwent an impressive radiation in the Southern Hemisphere, in both the South Atlantic and South Pacific. The northern fur seal (*Callorhinus ursinus*) evolved during the late Pliocene and was an important constituent of the Pacific Ocean marine mammal fauna.

Extant genera of both mysticetes and odontocetes appear during this time interval in the North Pacific (Alaska to California and Japan), North Atlantic (Canada-Champlain Sea, Florida, western Europe-Belgium). For example, bowheads evolved earlier during the Pliocene, but moved between the Atlantic and Pacific during warm interglacials when sea ice was reduced.

The last of the hydrodamaline lineage, which included the Steller's sea cow, flourished briefly in the cool waters of the North Pacific during the Plio-Pleistocene, ranging from Baja California to the Aleutians. Some high-latitude taxa were unable to disperse across the equator, resulting in various pinnipeds and cetaceans (for example, dusky dolphins and Pacific white-sided dolphins; see chapter 1 and fig. 1.5) confined to either the Northern or Southern Hemisphere, where they speciated. The rise and fall of sea levels during glacial and interglacial intervals likely contributed to speciation by reducing gene flow and isolating populations and is likely responsible for divergence of various species of *Phoca* in the Northern Hemisphere. By this time, the three living manatee species occupied coasts and rivers in Africa, South America, and the southeastern United States.

Sea otters (*Enhydra lutris*) evolved in the North Pacific at the beginning of the Pleistocene, about 1–3 Ma, and have not dispersed since that time. Polar bears (*Ursus maritimus*) are the most recent marine mammal species, originating between 1 and 1.5 Ma in the Arctic.

WHAT LED MARINE MAMMALS
BACK TO THE SEA?

This is a difficult question but likely involves feeding, since all major clades of marine mammals evolved morphological adaptations that allow them to efficiently process aquatic food. This makes sense when we consider that the origin and initial diversification of marine mammals were tied to increased temperatures and primary productivity that occurred in the Eocene. Extinct desmostylians and aquatic sloths, as well as dugongs and manatees, specialize on aquatic plants, whereas pinnipeds and whales feed on fish, squid, and vertebrate prey. Walruses and sea otters are invertebrate specialists, feeding on mollusks and urchins, respectively.

CHAPTER THREE

Pinniped Diversity, Evolution, and Adaptations

In this chapter, I trace the origin and evolutionary history of the major lineages of fossil and living pinnipeds and the major structural and functional innovations that shape their biology and behavior.

Pinnipeds, or "fin-footed" carnivores, are divided into three groups: seals (phocids), fur seals and sea lions (otariids), and walruses (odobenids). There are 34–36 extant species of pinnipeds, excluding the Caribbean monk seal (*Monachus tropicalis*) and the Japanese sea lion (*Zalophus japonicus*), which only recently went extinct. Both species vanished due to human activities. The largest pinnipeds are elephant seals, *Mirounga* species, with males reaching more than 5 m (16 ft) in length and weighing more than 2.5 tons (5,600 lb). Ringed seals (*Pusa hispida*), at a length of approximately 1.5 m (nearly 5 ft) and weighing up to 100 kg (220 lb), are among the smallest pinnipeds. Pinnipeds have a worldwide distribution and occupy a wide range of habitats from freshwater lakes to oceans. Relationships among the major lineages of pinnipeds are shown in fig. 3.1.

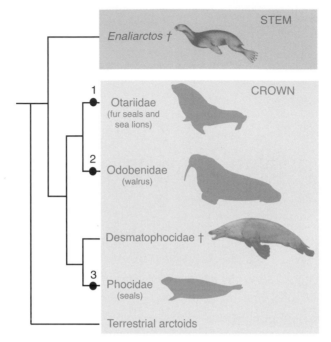

Figure 3.1. Evolutionary relationships among pinnipeds. Selected shared derived characters identified by numbers.

1. **Otariidae:** Frontal bones extend between the nasals; supraorbital process of the frontal bone located above the eye orbit; large and shelf-like; a ridge (secondary spine) subdivides the supraspinous fossa of scapula; pelage units (a primary hair and its surrounding secondaries) are spaced uniformly; trachea has an anterior bifurcation of the bronchii.
2. **Odobenidae:** Pterygoid strut; horizontally positioned expanse of bones lateral to the internal nares and hamular process; broad and thick calcaneum (heel bone) with prominent medial tuberosity; enlarged upper canine tusks in living species.
3. **Phocidae:** Lack the ability to draw hind limbs forward under the body; thick, dense (pachyostotic) ear region; supraorbital process of the frontal bone absent; strongly everted ilia (pelvic bone).

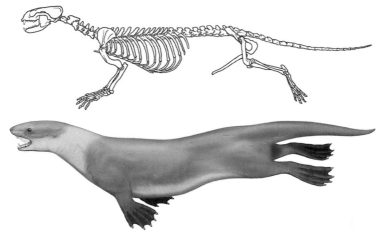

Figure 3.2. Skeletal and life reconstruction of *Puijila darwini* (painted by Carl Buell).

THE EARLIEST PINNIPEDS: WEBBED FEET OR FLIPPERS?

Pinnipeds shared a common origin probably during the Late Oligocene (27–25 Ma). Most scientists now agree that the earliest pinnipeds were descended from either bears (ursids) or otter-like forms (mustelids) known as arctoid carnivorans.

The recent discovery of a semiaquatic carnivore, *Puijila darwini*, from 20–24 Ma rocks in the Arctic, has been suggested as a morphological intermediate in the land-to-sea transition of pinnipeds (fig. 3.2). Unlike pinnipeds, it did not have flippers and more nearly resembles otters in its possession of a long tail and large, probably webbed feet. Whether *Puijila* represents an early pinnipedimorph or an early arctoid more distantly related to pinnipeds requires more study.

The best known early stem pinniped was *Enaliarctos*, which includes four species. Known from the middle Miocene (27–25 Ma) of the North Pacific, *Enaliarctos* possessed **heterodont** teeth with large blade-like cusps on the upper teeth, well adapted for shearing. Other pinniped species show a trend toward decreasing shearing function of the cheek teeth. These modifications resulted later in the development of the simple, peg-like or **homodont** teeth characteristic of most living pinnipeds. *Enaliarctos mealsi*, known by a nearly complete skeleton, was an animal similar in size and weight to a modern harbor seal. It was approximately 1.4–1.5 m (4.6–4.9 ft) in length and weighed between 73–88 kg (160–194 lb). Both the forelimbs and hind limbs of this animal were modified as flippers and used in swimming. However, the long lumbar region, larger hind limbs, and laterally bent pelvic bone suggest that *Enaliarctos* may have been a hind limb–dominated swimmer. Several features of the skeleton suggest that *Enaliarctos* was also highly capable of maneuvering on land and likely spent more time near the shore than extant pinnipeds.

CROWN PINNIPEDS

Crown pinnipeds include three extant families: Otariidae, Odobenidae, and Phocidae. Molecular and combined (morphological and molecular) data consistently support a closer alliance between otariids and odobenids (see fig. 3.1).

Otariidae (Fur Seals and Sea Lions)

Otariids are characterized by the presence of external ear flaps, and for this reason they are sometimes called eared seals. Another characteristic of otariids that can be used to distinguish them from phocids is their method of locomotion on land. Otariids (and odobenids) can turn their hind flippers forward and use them to walk on land, but phocids are unable to do so. Otariids generally are smaller than most phocids and are shallow divers, targeting fast-swimming fish as their major food source.

Otariids include two groups: fur seals and sea lions. Fur seals are named for their thick, dense fur; sea lions have relatively sparse pelage but a thick layer of blubber as an insulator (fig. 3.3). Although it is

Figure 3.3. Representative otariids: (a) California sea lion (*Zalophus californianus*) and (b) northern fur seal (*Callorhinus ursinus*) and (c) walrus (*Odobenus rosmarus*). Painted by Carl Buell.

generally accepted that northern fur seals (*Callorhinus ursinus*), which occur only in the North Pacific, are sister to remaining otariids, relationships among other members of this group are not well resolved. It now appears that some species of southern fur seals, *Arctocephalus* and *Arctophoca*, are more closely related to various sea lions than they are to other fur seals. Thus the traditionally accepted subfamilies Otariinae (sea lions) and Arctocephalinae (fur seals) are not valid.

The earliest known otariid is *Pithanotaria starri* from the late Miocene (11 Ma) of California. It was a small-bodied animal characterized by double-rooted cheek teeth and a fur seal–like skeleton. Another larger otariid appearing slightly later (8–6 Ma) is *Thalassoleon*, known from California, Mexico, and Japan. Features of the skull and teeth indicate that *Thalassoleon* was likely a fish eater and employed pierce feeding (biting) like most modern otariids. Also like modern otariids, *Thalassoleon* was a strong forelimb swimmer but differed in being better adapted to moving on land.

Otariids originated in the North Pacific but underwent a major diversification in the Southern Hemisphere once they crossed the equator in the last 2–3 Ma.

Odobenidae (Walruses)

A single species of extant walrus (*Odobenus rosmarus*) exists today and is readily identified by its elongate upper canine tusks developed in both males and females (fig. 3.3). Walruses today occupy Arctic and Subarctic regions. Two subspecies of walrus are recognized on the basis of geographic, morphologic, and molecular differences: *Odobenus rosmarus rosmarus* in the North Atlantic and *Odobenus rosmarus divergens* in the North Pacific.

Walruses first appear in the North Pacific during the Miocene (16–14 Ma). Walruses were considerably more diverse in the past, with at least 20 described fossil species in 14 genera distributed along the Pacific coast from Japan to Baja California. Stem walruses (for example, *Proneotherium*, *Neotherium*, and *Imagotaria*) are arranged as sequential sister taxa to later diverging clades, dusignathines, and odobenines. Skulls and partial skeletal remains of the extinct stem walrus *Proneotherium repenningi* show evidence of aquatic limb adaptations, including a flexible lower spine, short hind limbs, and paddle-shaped feet. Interestingly, the fossil record indicates that "tusks do not a walrus make." Dusignathine walruses developed enlarged upper and lower canines, whereas only odobenines evolved the large upper canines seen in the modern walrus. Among the more interesting fossil odobenine walruses is *Valenictus chulavistensis*, which lacks

Figure 3.4. Representative monachine and phocine seals: (a) the monachine northern elephant seal (*Mirounga augustirostris*) and (b) the phocine ribbon seal (*Histriophoca fasciata*; painted by Carl Buell).

teeth in both jaws with the exception of two tusks in the upper jaw. Such extreme toothlessness evolved convergently in the narwhal.

Phocidae (Seals)

The second major grouping of living seals, the phocids, are often referred to as the earless seals for their lack of visible ear pinnae, a characteristic that readily distinguishes them from otariids. Another characteristic is their method of movement on land. Phocids are unable to turn their hind flippers forward because of the peculiar structure of their ankles, and progression on land is accomplished by undulations of the body rather than movement of the hind flippers or foreflippers. Other characteristics of phocids include their large body size in comparison to otariids. Several phocids, most notably elephant seals and Weddell seals, are spectacular divers that feed on squid and fish at depths of 1,000 m (0.5 mi) or more.

Most pinnipeds (19 species) are phocid seals. Although traditionally phocids have been subdivided into as many as four major subgroups, there is general consensus that only two of these subgroups are valid (that is, monophyletic): the monachines or Southern Hemisphere seals (including monk, elephant, and Antarctic seals) and the phocines or Northern Hemisphere seals (10 species inhabit the Arctic and Subarctic; fig. 3.4).

There is an earlier, less documented record of phocids from the late

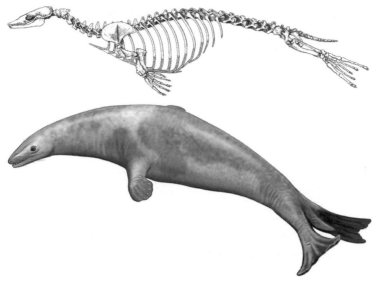

Figure 3.5. Skeleton of a stem phocid (*Acrophoca longirostris*) from the Miocene of Peru (from Muizon 1981) and life restoration (painted by Carl Buell).

Oligocene of South Carolina. Undisputed evidence for both monachine and phocine lineages on both sides of the Atlantic is from the middle Miocene (approximately 15 Ma). Stem monachines include *Monotherium, Leptophoca,* and *Callophoca. Leptophoca lenis* is well represented by skulls and several nearly complete skeletons from the North Atlantic that exhibit **sexual dimorphism**, a common characteristic of some pinnipeds. *Leptophoca* was similar in size to a modern harp seal (the latter ranges between 1.6 and 1.7 m or between 5.2 and 5.5 ft in length, weighing 130 kg or 287 lb). Features of the hind limbs, similar to those of modern seals, indicate that this stem monachine seal was a hind limb swimmer.

Stem phocines include *Acrophoca, Piscophoca,* and *Homiphoca. Acrophoca longirostris,* known from late Miocene and/or early Pliocene South America, is unique among phocids in its long, slender, flexible neck and elongated body (fig. 3.5). The everted pelvis and modified hind flippers of *Acrophoca* suggest that it was likely a hind limb–dominated

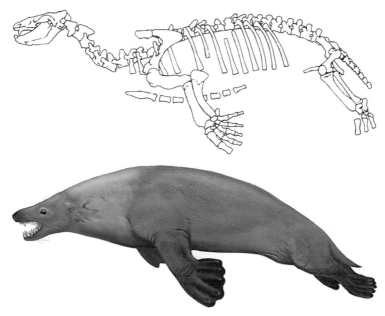

Figure 3.6. Skeleton of desmatophocid (*Allodesmus kernensis*) from the Miocene of western North America; original 2.2 m long (see Berta et al. 2006 for original source) and life restoration (painted by Carl Buell).

swimmer. *Acrophoca* was approximately 1.5 m (nearly 5 ft) in length. Its diet probably consisted primarily of fish.

DESMATOPHOCIDS:
EXTINCT PHOCID RELATIVES

Desmatophocids (*Allodesmus* and *Desmatophoca*) are an extinct lineage of pinnipeds known from the early middle Miocene (23–15 Ma) of the North Pacific. They are characterized by pronounced sexual dimorphism. *Allodesmus* appears to have been similar in size to some of the larger crown pinnipeds but, like *Enaliarctos*, seems less aquatic. Large forelimbs and a short lumbar region provide evidence that *Allodesmus* was a forelimb-dominated swimmer (fig. 3.6). The currently accepted hypothesis links desmatophocids with phocid pinnipeds on the basis of several skull characters.

EVOLUTIONARY TRENDS

The earliest well-documented pinnipedimorphs (stem and crown pinnipeds) were moderate-sized animals (about the size of a harbor seal) with well-developed flippers and a small tail. If *Puijila* is a pinnipedimorph instead of an arctoid, then a webbed feet stage is represented before the transition to flippers. Later in their evolution, some desmatophocid pinnipeds evolved a larger body size and sexual dimorphism and moved from coastal regions to forage in oceanic habits, although they still had to return to land annually to breed and molt. The relatively larger body size of some phocids, such as elephant seals and some extinct desmatophocids, has been related to their large oxygen stores and deep diving capabilities, discussed later in this chapter. Postcranial remains suggest that the earliest pinnipedimorphs, such as *Enaliarctos*, were likely hind limb–dominated swimmers. Otariids evolved forelimb swimming and phocids and odobenids retained hind limb–dominated swimming, although the latter also employed the forelimbs. Some desmatophocids, notably *Allodesmus*, were more similar to forelimb-swimming otariids.

Early pinnipedimorphs, such as *Enaliarctos* and the walrus *Proneotherium*, likely fed on fish close to shore, employing shearing teeth capable of processing food in the mouth. Later pinnipeds (extinct otariids such as *Thalassoleon* and extinct walruses such as *Neotherium*) transitioned to more of a piercing dentition that served primarily to seize and hold prey, which was then swallowed whole. Other pinnipeds evolved specialist strategies—most notably, benthic suction feeding among odobenine walruses and some otariids, such as the southern sea lion, and filter feeding in crabeater seals.

Body length:	50 cm	100 cm
Max. girth = G:	30 cm	60 cm
Surface area = 0.7 GL^2:	1050 cm^2	4200 cm^2
Volume = 0.02GL^3:	1500 cm^3	12000 cm^3
SA/V Ratio:	0.70:1	0.35:1

Figure 3.7. Surface-to-volume relationship. As a body increases in size (left to right), its surface area and volume also increase; however, the ratio of its surface area to volume decreases.

STRUCTURAL AND FUNCTIONAL INNOVATIONS AND ADAPTATIONS
Surface Area and Volume

Since water conducts heat 25 times more efficiently than air, heat flows out of a warm body in cold water more rapidly than it does when the same body is in air. The ratio of **surface area** to **volume** is an important concept that governs the body size of an animal. The rate of heat loss depends on surface area. Therefore, to stay warm an animal needs an effective insulator (for example hair, blubber) and a small surface (think small flippers). As an animal gets larger, it has progressively less surface area in relation to its volume (fig. 3.7). Most marine mammals are large and streamlined with few, small, projecting appendages and they are thus capable of producing considerable heat with relatively little loss at the surface.

Blubber versus Fur

Typically, otariids have fur, with fur seals having the densest coat, whereas phocids rely on **blubber**, a loose connective tissue containing fat cells for insulation. Because fur traps air, it is a good insulator and grooming ensures that it is maintained. Some newborn seals, such as harp seals, rely on **lanugo**, or pup fur, which is both long and fluffy and provides insulation when the pup is on land. A pup must shed lanugo and develop a substantial blubber layer before entering the water. Blubber is also a good insulator, as well as a primary source of metabolic fuel, and plays a role in buoyancy regulation.

Like other mammals, pinnipeds must **molt** or shed their fur in order to maintain a healthy coat. Molting usually occurs once a year in the spring or fall after breeding, although sea lions and fur seals tend to molt their fur fairly gradually all year long. Fur seals may take up to three years to complete a molt. Some phocids (such as elephant seals and Hawaiian monk seals) go through an abrupt "catastrophic molt" and will actually lose patches of skin and hair in sheets all at once.

Color Patterns: Concealment and Recognition

The color of a pinniped (or any marine mammal) is determined by pigment cells in the epidermis. The most widespread pattern is **countershading**, in which the animal has a dark back (dorsal surface) and a light underside (ventral surface) such as Weddell seals, which eliminates any shadows that might result from top-lit areas of the ocean and make the animal conspicuous to predators. Some seals, such as ringed and ribbon seals (fig. 3.4), show contrasting light and dark disruptive patterns. Such distinctive markings serve for individual recognition. A uniform color pattern, such as the completely white coloration of harp seal pups, serves to hide them from predators and keep them warm until they build up a blubber layer and develop a darker pelt.

Moving on Land and in the Water

THE PINNIPED FLIPPER, SKELETON, AND LOCOMOTION

Pinnipeds, unlike cetaceans and sirenians, have pairs of flippers on both the forelimbs and hind limbs. The limbs are enclosed within the body profile to the level of the elbow and ankle (fig. 3.8). They are used mainly

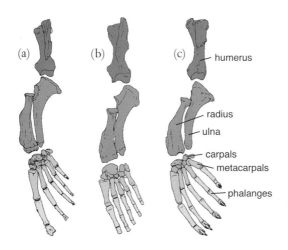

Figure 3.8. Comparison of the forelimbs of pinnipeds: (a) sea lion, (b) walrus, and (c) harbor seal (see Berta et al. 2006 for original data sources). Shading denotes upper arm, forearm, and hand.

for locomotion in the water. Otariids use well-developed muscles of their foreflippers to increase thrust underwater, with the hind limbs used in steering or trailing passively behind. The forelimb bones of otariids are shortened relative to terrestrial carnivores. This shortening of the resistance arm of the forelimb increases the mechanical advantage of the primary locomotor muscles rotating at the shoulder and elbow joints. In this way, sea lions and fur seals "fly" through the water resembling penguins and sea turtles (fig. 3.9). Conversely, the forelimbs of phocids are used solely in steering, with alternate sweeps of the more powerful hind limb musculature providing propulsion. Walruses use the forelimbs as paddles, with most of the propulsion coming from the hind flippers (fig. 3.9). The foreflipper and hind flipper of nearly all pinnipeds are distinguished by having the fingers and toes, or digits, elongated by the development of extensions of cartilage at the end of each digit. Phocids are incapable of turning the hind limbs forward due to the structure of the ankle and heel bones, so movement on land is accomplished by undulations of the body. Otariids and the walrus are able to turn their hind flippers forward and "walk" on land.

In otariids, foreflipper propulsion on land is aided by extensive head and neck movements facilitated by relatively large neck vertebrae. In

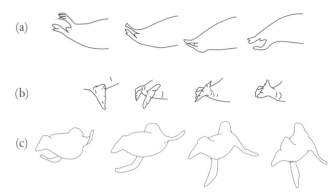

Figure 3.9. Comparison of aquatic locomotion in pinnipeds: (a) phocid, (b) walrus, and (c) otariid; see Berta et al. 2006 for original data sources).

walruses or phocids, the lower back (lumbar) vertebrae have long horizontal processes that provide a greater surface area for muscles that move the posterior end of the body.

SWIMMING, SPEED, AND THE COST OF SWIMMING

Since water is denser and more viscous than air, it also creates resistance against forward movement or **thrust**. Resistance, or **drag**, develops on the surface of the skin as the animal moves in water. To minimize drag, the body has become smooth-surfaced and streamlined with the elimination of projecting body parts. Sea lions move forward using foreflippers to generate propulsion and create **lift**. Seals and walruses rely on hind flippers for forward propulsion (thrust) and lift (fig. 3.10).

Swimming is the most energetically efficient form of locomotion, since swimmers do not need to support their weight as do flyers and land mammals. The most efficient swimming speed depends on anatomical characters of the flipper, such as size and shape. Foreflipper propulsion provides several advantages, including stability at slow speeds and maneuverability at high speeds. Consequently, otariids are champion underwater acrobats and are capable of rapid changes in direction and speed. Otariids routinely swim at speeds of 2–3 m or 6.5–9.8 ft/sec, and

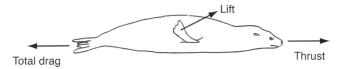

Figure 3.10. Lateral view of sea lion to show drag, lift, and thrust.

phocids, swimming at average speeds, proceed more slowly at 1.2–2.0 m or 3.9– 6.5 ft /sec. In comparison, world-class human swimmers reach average speeds of 1.82 m or nearly 6 ft/sec.

FLIPPERS AS THERMOREGULATORS

Flippers also serve as **thermoregulators**. Pinnipeds and other marine mammals alter blood flow to flippers according to ambient conditions, effectively using them as a radiator to dump heat. The flippers (and dorsal fin of some whales and tail flukes of whales and sirenians) contain an arrangement of blood vessels, termed **countercurrent**, in which an artery containing warm blood from the body core is surrounded by a bundle of veins containing cool blood from the body extremities (fig. 3.11). As the blood in these vessels flows past each other, a heat gradient is created in which heat is transferred. The warm arterial blood transfers most of its heat to the cool venous blood in the flippers. This conserves heat by recirculating it back to the body core. Since the arteries give up a good deal of their heat in this countercurrent exchange, there is less heat lost at the surface of the flipper, dorsal fin, or tail.

UNDERWATER SIGHTS AND SOUNDS

Pinnipeds have the largest number of **vibrissae** (whiskers) of any mammal. Vibrissae are tactile hairs. The most prominent whiskers are usually found on the snout (mystacial vibrissae). Vibrissae are usually longer and stiffer than body hairs and their follicles are surrounded by complex blood sinuses and mechanoreceptors. Research on various pinniped species, including harbor and ringed seals, California

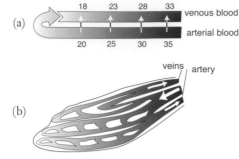

Figure 3.11. General pattern of heat exchange in (a) countercurrent system and (b) simplified heat exchange network of a pinniped flipper.

sea lions, and walruses, has shown that their well-developed vibrissae contain nerve fibers that serve as sensitive detection systems, enabling them to precisely locate and track prey. In fact, study of the tracking ability of seal whiskers indicates that they are as good at detecting prey as are echolocating dolphins (see chapter 4). Experiments have shown that ringed seal vibrissae are sensitive to waterborne sound waves and they use this information to navigate in the dark. Other seals use vibration detected by their whiskers to follow fish "trails" in the water in order to capture and eat them.

The eyes of most pinnipeds are large relative to body size. The walrus is an exception with its small, laterally positioned eyes. In basic structure, the eye of a pinniped is like that of other vertebrates, but the shape of the lens is spherical. When out of water, the spherical lens makes vision nearsighted, although this is partly compensated for by smaller pupils. The retina is dominated by rods that provide increased sensitivity to low light. Some form of color vision is likely although it appears limited to certain light conditions. Glands located at the outer corners of the eyes produce an oily mucous that protects the eye and cornea from drying out.

Pinnipeds produce airborne and underwater sounds. Airborne sounds are usually within the range of human hearing and are described as grunts, snorts, or barks. They are often identified with their presumed social function, such as "threat calls" of breeding males or "pup attraction calls" of mothers. Most pinniped vocalizations are produced in the "voice box" or larynx, although male walruses also make clacking noises with their teeth and produce distinctive bell-like sounds in air and underwater with their inflated pharyngeal (throat) pouches.

Diet and Feeding Specializations

Since pinnipeds, for the most part, swallow prey (mostly fish and squid) whole, the cheek teeth are typically **homodont**. A notable exception is the walrus, with a pair of upper canines enlarged as tusks. Both males and females possess tusks, which are not usually used in feeding but rather in dominance displays and to pull themselves onto ice floes.

Walruses differ from most other pinnipeds by suction feeding and they typically feed on clams. They swim along the sea bottom in a head-down position, using their sensitive vibrissae to locate prey. When a clam is located, it is excavated with a powerful jet of water squirted from the mouth. The meat of the clam is then sucked out of the shell by the tongue. Walruses have been reported to consume as many as 6,000 clams per meal! Anatomical adaptations for suction feeding include an arched palate, reduced teeth, and a tongue that works like a piston creating low pressure in the mouth.

Walruses were apparently not always proficient suction feeders. Fossil walruses that lived during the Miocene (that is, *Imagotaria* and *Neotherium*) possessed teeth but no tusks and had a relatively flat palate similar to that of their fish-eating relatives, otariid seals. It has been suggested that this shift in feeding strategy among walruses took place at least 11 Ma. Other pinniped specialist feeders are the leopard and crabeater seals. The leopard seal (*Hydrurga leptonyx*) is a grip-and-tear (a type of biting) feeder that possesses enlarged front teeth and sharp cheek teeth and feeds on birds and other marine mammals. The crabeater seal (*Lobodon carcinophagus*) has teeth with intricate cusps that function as sieves to filter **krill**.

RESEARCH TOOLS: TIME-DEPTH
RECORDERS AND CRITTERCAMS

Researchers have outfitted elephant seals with satellite-based tracking devices and **geographic location time depth recorders** (GLTDRs) that provide details of their diving behavior (fig. 3.12). Recent studies equipped with video cameras or **crittercams** have provided amazing insights into behavior, such as Weddell seals flushing prey from crevices in the ice by blowing bubbles.

Our understanding of pinniped diets has also been improved through the advent of several new techniques. DNA markers have been used to determine individual, sex, and species identification of prey in pinniped fecal samples. Lipids such as fatty acids can also provide very specific information about the diet of pinnipeds and other marine mammals. Comparison of reference **fatty acid signatures** of various prey species to the fatty acid signature of the predator obtained from blubber or milk

Figure 3.12. Harbor seal with TDR glued to skin and rear flipper tags (courtesy B. Stewart, Hubbs–Sea World Research Institute).

reveals diet composition. Stable isotope ratios of carbon and nitrogen found in various pinniped tissues (for example, muscle, blood) reflect prey species eaten over a period of time and can provide information on foraging location and trophic level (see also chapters 1 and 6).

INCREDIBLE DIVING MACHINES:
WEDDELL AND ELEPHANT SEALS

Many marine mammals are capable of prolonged and deep dives. Northern elephant seals hold the record for pinnipeds, an ability to dive for more than one hour at a depth of over 1,600 m (1 mi) on a single breath of air. Another phocid, the Weddell seal in the Antarctic, is also an extreme diver.

Numerous adaptations of the respiratory system facilitate deep diving, such as flexible ribs that allow the lungs to collapse. As air is forced out of the lungs, the compressed air in the larger air passages is blocked from contact with the thin, gas-exchanging walls of the alveoli. As a result, little compressed gas is absorbed by the blood during dives. Marine mammals thus avoid the problems of decompression sickness (also called the **bends**) that sometimes plague human divers breathing compressed air. Oxygen is stored in the muscles and the blood rather than in the lungs as in human divers. Deep-diving pinnipeds and whales possess high concentrations of **myoglobin**, a protein pigment found in muscle cells that gives them a dark red appearance, as well as hemoglobin in red blood cells. Myoglobin functions as an oxygen storage site, providing oxygen to the muscles. Pinnipeds usually exhale before diving and close their nostrils and throat cartilages, preventing water from entering the windpipe or trachea. They remain submerged with empty lungs.

The long dives of these animals are accomplished by a variety of circulatory changes, including a slowed heart rate, reduced oxygen consumption, and shunting blood only to essential organs and tissues. Oxygen storage sites in marine mammals differ from those in human divers in that a larger volume of oxygen is stored in the muscles and blood relative to that stored in the lungs (fig. 3.13).

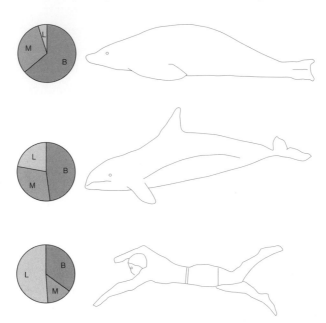

Figure 3.13. Generalized comparison of relative blood (B), muscle (M), and lung (L) stores for pinnipeds, odontocete cetaceans, and humans (see Berta et al. 2006 for original data source).

The diving patterns of pinnipeds vary. Body size is an important factor since larger body size reflects larger oxygen storage capabilities. Otariids do not spend as much time at sea diving as phocids and they usually dive for only a few minutes at relatively shallow depths. Otariids have relatively small, sleek bodies consistent with an "energy release" predator lifestyle. Some larger otariids are exceptions, such as the New Zealand sea lion, a relatively deep diver. Although walruses are among the largest pinnipeds, they normally dive for short periods to very shallow depths. This reflects the fact that they feed in shallow water on mollusks. However, given their large body sizes, walruses are capable of diving to much greater depths than those recorded. Phocids, by comparison, employ an "energy conservative" strategy, remaining at sea for long periods during feeding bouts. Again, there is variability

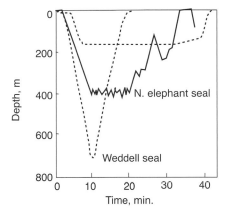

Figure 3.14. Time-depth profiles of Weddell and northern elephant seals (see Berta et al. 2006 for original data sources).

in body size among phocids. Small species such as Baikal, ringed, and harbor seals usually dive for only a few minutes to relatively shallow depths, while the much larger elephant seal has been dubbed an incredible diving machine for its prodigious diving capabilities.

Weddell seals perform two main types of dives: short, hunting dives and long, exploratory dives (fig. 3.14). Short dives to locate food last up to 20 minutes and require no major heart rate changes. These dives appear to be within the **aerobic dive limit** (**ADL**) of the species, which is defined as the longest dive that does not lead to an increase in lactic acid concentrations during the dive. Some dives, employed when the seal is in search of new breathing holes in the ice, last longer than 20 minutes. These longer dives are characterized by adjustments to the respiratory and circulatory systems, and result in an increase in blood lactate concentration, which signifies that an animal is operating under anaerobic rather than aerobic metabolism. This is similar to the experience of humans after hard exercise, when lactic acid builds up in the muscles, but deep-diving marine mammals are able to tolerate large lactic acid accumulations and perform anaerobically for longer periods of time than humans. These depleted oxygen stores, however, must be replenished and aerobic metabolism restored by a recovery time spent at the surface. After long dives, Weddell seals are exhausted and sleep for several hours.

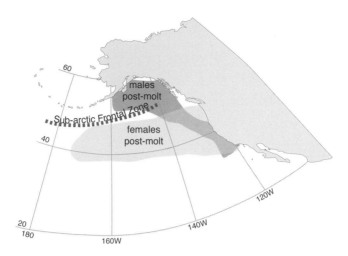

Figure 3.15. Geographical distribution of male and female northern elephant seals during postmolt foraging patterns (from Berta et al. 2006).

Elephant seals employ a different diving strategy. Both sexes dive day and night for 20–30 minutes per dive without prolonged rest periods at the surface (fig. 3.14). The short duration of these dives suggests that elephant seals are diving within their ADL. They remain at sea for several months during two feeding migrations. These yearly double migrations of 18,000–21,000 km (11,000–13,000 mi; fig. 3.15) rival those of gray and humpback whales (see next chapter). Northern elephant seal females forage offshore along the Pacific coast. Adult males travel further north, feeding on large squid in the rich areas of productivity located in the Aleutian Islands and the Gulf of Alaska.

MATING AND SOCIAL SYSTEMS, REPRODUCTION, AND LIFE HISTORY

In the past 20 years, there has been an enormous increase in the number and quality of field studies on the ecology and behavior of pinnipeds. In addition to satellite telemetry, time-depth recorders, and crittercams

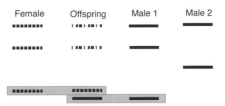

Figure 3.16. Representation of paternity testing by DNA fingerprinting (see Berta et al. 2006 for original data source). Offspring bands were derived from the mother (even dashes), the father (solid lines), or both (uneven dashes). Male 2 may be eliminated as the father because he and the offspring share no bands.

(described earlier), advances in molecular biology have provided unparalleled opportunities to examine interactions between individuals in populations to determine paternity and kinship. For example, **DNA fingerprinting** (the name comes from the analogy to fingerprints) can be used to determine parentage, since approximately 50 percent of the DNA profile (genetic loci represented as bands on a gel) in the offspring come from the mother and 50 percent come from the father. Thus it is possible to exclude individuals as parents if the bands in the offspring are not present in the fingerprint of either parent (fig. 3.16).

Courtship: Inflated Noses and Hoods

Courtship patterns are important for species recognition. Males of some species have elaborate **secondary sexual characters** that are used during courtship. For example, male elephant seals possess an enlarged nose. When inflated, it hangs down in front of the mouth. During dominance battles with other males, snorts and other vocalizations are deflected downward into the open mouth, which acts as a resonating chamber. Male hooded seals display an inflatable hood, an enlargement of the nasal cavity on top of the head, to attract females. They also possess the ability to blow a red, balloon-shaped structure from their nostril during mating season.

Polygyny, Monogamy, and Leks

Pinniped mating systems range from **monogamy** in which both sexes mate with one partner, to the most extreme **polygyny** where a male has multiple mates during a breeding season. When an individual male can control or gain access to several females, the male can increase his **reproductive success** by mating with more than one female. In male dominance polygyny, males compete for females by establishing dominance over other males or by defending breeding territories occupied by females. Polygyny is more common among pinnipeds that mate on land, which is approximately one-half of the species.

Elephant seals are a familiar example of male dominance polygyny. In addition to their larger size, male elephant seals have additional secondary sexual characteristics, including an inflatable nose, enlarged canine teeth, and thick skin on the head and neck. Breeding rookeries are normally located on islands or isolated mainland beaches. Consequently, rookery space is limited and females are crowded. Males establish dominance hierarchies within aggregations of females. The most dominant (alpha) male in a hierarchy defends nearby females from subordinate males. At Año Nuevo Island on the central California coast, northern elephant seal breeding behavior has been studied for more than three decades by Burney Le Boeuf and his colleagues. According to their research, fewer than 10 percent of males manage to mate at all during their lifetimes, whereas very successful males mate with 100 or more females.

Otariid seals, such as northern fur seals and Steller sea lions, exhibit resource defense polygyny, in which males establish and defend specific breeding territories. Territorial defense behavior consists of male vocalizations and threat displays rather than physical combat, as seen in elephant seals.

Aggregations of many males in a small area for purposes of mating display, called **leks**, are seen in Pacific walruses and some ice breeding seals. During the breeding season, male walruses position themselves near female haul outs and perform underwater displays, producing a series of bell- and gong-like sounds to attract females. Male Weddell, ringed, and bearded seals are also known or suspected to defend underwater territories near female haul outs and to perform vocal displays. The females presumably choose males based on the quality of the territories and displays.

Those ice breeding pinnipeds that mate in the water and give birth on ice include Weddell, leopard, Ross, and crabeater seals in the southern oceans and spotted, ringed, ribbon, bearded, hooded, and harp seals in the north. Since ice is an unstable habitat, females do not densely aggregate and extreme polygyny does not develop. Instead, nearly all ice breeding pinnipeds are monogamous or slightly polygynous. The extensive distribution of sea ice promotes dispersal of females and, as a result, it is difficult for males to monopolize several females. Several ice breeding pinniped males exhibit reversed sexual dimorphism. For example, Weddell seal males are actually smaller than females, which has been suggested to increase their swiftness in the water, therefore making them more attractive as a potential mate.

Postpartum Estrus and Delayed Implantation

The reproductive organs of pinnipeds (and other marine mammals) are for the most part hidden beneath the skin, contributing to their streamlined bodies. Male pinnipeds, in addition to sea otters and polar bears, have a bone in their penis, the os penis or **baculum**. All pinniped females except the walrus have a **postpartum estrus**, in which estrus or sexual receptivity closely follows parturition or birth. This ensures that mating occurs while both sexes are together on land or ice as opposed to being at sea during feeding migrations. The unpredictable nature of the location, extent, and breakup time of pack ice in particular restricts the breeding season to a short period of time.

An important reproductive feature of pinnipeds, as well as of sea otters and polar bears, is **delayed implantation**, the ability to maintain the embryo in a state of suspended development for a period of a few weeks to several months prior to implantation in the uterus. This enables the mother to separate the time of mating and fertilization from the start of gestation, the time from fertilization to birth. Thus delayed implantation allows mating and birth of the young to occur at optimal times of the year, such as spring and summer, when food availability is best.

Life History

SEXUAL MATURITY

Some pinnipeds, particularly polygynous species, are characterized by **sexual bimaturity**, in which one sex matures before the other. Before polygynous males can compete successfully for breeding territories or mates, they must achieve a body size substantially larger than females and gain more experience fighting. For this reason, in male elephant seals sexual maturity occurs at six years of age whereas in females this occurs earlier, at approximately four years of age.

PREGNANCY AND BIRTH

The gestation period in pinnipeds is approximately one year when delayed implantation is included. A single offspring is typical of all marine mammals except polar bears. Typically, the young are born **precocial**, self-sufficient, and are capable of swimming and diving shortly after birth. But some species, such as the Australian sea lion, are **altricial**, or dependent on mom at birth. Pinnipeds typically give birth annually, although walruses are exceptions with a reproductive interval of four to six years.

TABLE 3.1

Lactation strategies in marine mammals.

Feature	Fasting	Foraging Cycle	Aquatic Nursing
Duration of fasting	All of lactation	Variable (a few days)	Short (hours–days)
Duration of lactation	Short (~4 weeks)	Intermediate (~4 months)	Long (~2–3 years)
Fat content of milk	High (55%)	Intermediate + (40%)	Long (20%)
Pups/calves forage during later lactation	No	No	No

PARENTAL CARE AND LACTATION STRATEGIES

All marine mammals produce fat-rich milk, ranging from 30 to 60 percent fat, compared to 4 percent fat in human milk! Pinnipeds exhibit three maternal care and lactation strategies that reflect their respective ecologies: (1) fasting, (2) foraging cycle, and (3) aquatic nursing (table 3.1). The **fasting strategy** characterizes some phocid seals, notably the hooded, elephant, and some grey seals. This strategy is characterized by mothers that remain out of water for the entire duration of a relatively short lactation period, ranging from four days in the hooded seal to four to five weeks in elephant seals. High-fat milk (as much as 60 percent fat) is transferred from the mother to the rapidly growing offspring. Studies of northern elephant seals reveal that pups gain an average of 4 kg per day prior to weaning. After pups are weaned, they gather on beaches to form **weaner pods**. Pups that steal milk from other nursing cows, known as superweaners, typically grow much fatter than other pups. Crowded rookery conditions also lead to mother–pup separations, and the fostering of orphaned pups, known as **allomothering**, is frequently observed.

The **foraging cycle strategy** is exemplified by otariids and some smaller phocid seals (for example, harbor seals). This strategy is characterized by mothers that fast only for a few days while nursing their pups and then leave their pups onshore while they forage at sea. The duration of the foraging trips depends on the distance of foraging areas and onfood abundance and ranges from one day to three weeks. Females then return to land to nurse their pups. Otariid milk usually has a lower fat content than that of phocids.

The walrus shows the third maternal care strategy, **aquatic nursing**, in which the young are fed at sea and mothers forage during nursing. Walrus calves go to sea on foraging trips with their mothers once they have learned to dive, at approximately five months of age. The length of lactation in walruses is up to two years, the longest among pinnipeds, suggesting that considerable time is needed to learn proficient suction feeding techniques.

GROWTH AND LIFE SPAN

Females, especially those of polygynous species, tend to live longer than males. Males fight hard and die young, and most do not survive even until the age of delayed sexual maturity. In general, the life span of pinnipeds varies between 25 and 30 years.

The age of marine mammals with teeth, such as pinnipeds, can be determined by counting annual growth layers, which are analogous to tree rings. This method assumes that annual layers are readily discernible and that the time interval represented by these layers can be independently verified, neither of which is often the case.

CHAPTER FOUR

Cetartiodactylan Diversity, Evolution, and Adaptations

The name Cetartiodactyla reflects new discoveries that have revealed that whales are nested within even-toed ungulates (artiodactyls) such as hippopotamuses, giraffes, and deer. Two major groups of crown whales, or Cetacea (from the Greek word *cetus* meaning *whale*), are recognized: toothed whales, or Odontoceti, and baleen whales, or Mysticeti. Toothed whales are considerably more diverse, with approximately 73 known species, compared to 14 mysticete species. Whales display considerable diversity in size (fig. 4.1), ranging from the blue whale, the largest animal on Earth (33 m or more than 100 ft in length, weighing 150 tons) to the vaquita (1.4 m or 4.6 ft in length, weighing 42 kg or 92 lb).

Relationships among extinct and living whales and their close relatives are shown in fig. 4.2. Although whales and artiodactyls are linked, it is less certain which specific artiodactyl group is their closest kin. Molecular data points toward hippos but there is a significant gap in the fossil record of nearly 40 Ma from the earliest appearance of whales to that of hippos. Recent discoveries in Pakistan and western India of an extinct group of artiodactyls, the raoellids, have suggested that they are the closest extinct whale relatives (fig. 4.2). Raoellids lived during the early and middle Eocene (55–45 Ma). Raoellids are best represented

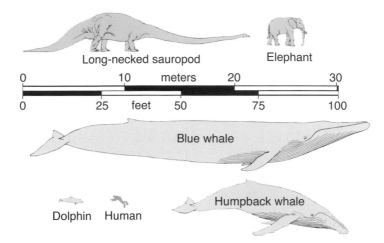

Figure 4.1. Size comparisons among various mammals and a dinosaur.

by *Indohyus*, a cat-sized animal with a long snout, long tail, and long slender limbs (fig. 4.3). Raoellids had very thick limb bones (**osteosclerosis**), an adaptation for buoyancy control (see chapter 5). At the end of each limb were four to five toes that ended in hooves, similar to those of a deer. The ankle bone of raoellids consisted of a double pulley similar to that seen in other artiodactyls and extinct cetaceans. This shape of the ankle bone creates greater anteroposterior mobility in the foot. Since raoellids were largely aquatic, this indicates that an aquatic lifestyle arose before cetaceans evolved. Most likely, raoellids were herbivores or possibly omnivores, consuming a mixture of plants and invertebrates.

EARLY WHALES HAD LEGS!

The earliest stem whales (Pakicetidae, Ambulocetidae, and Remingtonocetidae) are all known from the early and middle Eocene (50 Ma) of India and Pakistan (fig. 4.2). All are thought to have been semiaquatic, able to move on land as well as in the water. Their varied body forms

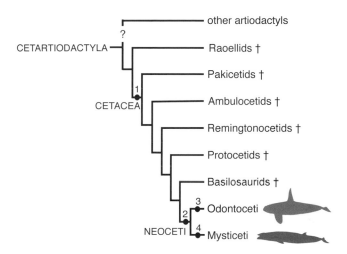

Figure 4.2. Evolutionary relationships among cetaceans. Selected shared derived characters of the groups identified by numbers.

1. **Cetacea:** Thick, dense (pachyostotic) ear bones of compact bone.
2. **Neoceti (Crown Cetacea):** Telescoping of skull roofing bones (see fig 4.4); fixed elbow joint.
3. **Odontoceti:** Nasal bones elevated above the rostrum; single nasal passage (except for physeterids); frontal bones higher than nasal bones; telescoping of skull in which maxillary bone covers the supraorbital process of frontals; presence of melon; cranial and facial asymmetry; echolocation (see also fig. 4.24).
4. **Mysticeti:** Lateral margins of maxillary bone thin; telescoping of skull in which the maxilla extends posteriorly under the frontal; unfused, ligamentous connection between lower jaws; laterally curved lower jaw; baleen present in all living taxa.

and number of species indicate considerable diversity during the transition from land to sea.

Pakicetids include *Pakicetus, Icthyolestes,* and *Nalacetus.* They show little resemblance to extant whales with their wolf-like proportions, long noses, and tails. Discoveries of pakicetid skeletons indicate that they had

Figure 4.3. Skeletal reconstruction of raoellid *Indohyus*. Hatched elements are reconstructed on the basis of related taxa (from Thewissen et al. 2007) and life restoration (painted by Carl Buell).

well-developed forelimbs and hind limbs. Pakicetids were mostly freshwater animals and are believed to have waded and walked in freshwater streams. Wear on their teeth is consistent with a fish-eating habit.

Ambulocetids are best known by *Ambulocetus natans*, which had a long, snouted skull similar to that of pakicetids, except that the eyes faced upward rather than to the sides of the head. They were much larger than any pakicetid, approximately the size of a large male sea lion. The limb proportions of *Ambulocetus* are similar to those of a river otter, and it has been suggested that they swam in a similar manner with their hind limbs and tail.

Remingtonocetids, known by four to five genera, are crocodile-like animals characterized by long, narrow skulls and jaws and robust limbs. Their molars show that they have lost the crushing specialization of pakicetids and ambulocetids, which suggests a difference in diet from earlier cetaceans. Like ambulocetids, they are also found in nearshore coastal settings, indicating a greater reliance on marine prey. The ear of remingtonocetids appears more specialized, as is the development of a large opening in the lower jaw, the mandibular foramen. In modern whales, the mandibular foramen contains a pad of fat that connects the lower jaw to the middle ear and provides a pathway for transmission of underwater sounds. This parallel structure in remingtonocetids suggests that they had evolved an ability for underwater hearing.

The occurrence of later diverging semiaquatic protocetids (fig. 4.2) in Asia, Africa, Europe, and North America indicates that cetaceans had spread across the globe between 49 and 40 Ma. Protocetids are diverse and well represented (15 genera). They differed from other early cetaceans in having large eyes with the nasal opening, which had migrated further posteriorly on the skull. Their heavy jaws and large teeth suggest a diet composed of fish or other vertebrates. Skeletons reveal that some protocetids, such as *Rodhocetus*, were able to move on land and in the water. Swimming was likely a combination of paddling with the hind limbs and dorsoventral undulations of the tail. The discovery of a near-term "fetus" in a protocetid skeleton (*Maicetus*) revealed positioning for head-first delivery, typical of land mammals but not whales, and the suggestion was made that early whales gave birth on land. However, it has been argued, based on necropsies of extant whales that have stranded, that the preserved smaller individual inside the body cavity may instead represent a displaced fetus or a prey item in the digestive tract of the larger individual.

Basilosaurids, recognized as the closest relatives of crown cetaceans (Neoceti; fig. 4.2) included both large and small body forms. They were widely distributed and lived between 41 and 35 Ma. Among the large-bodied forms is *Basilosaurus* (see chapter 2), which had a snake-like body consisting of elongate vertebrae with a maximum length of 17 m (56 ft). Best known is *Basilosaurus isis*, with several hundred skeletons reported from the middle Eocene Valley of Whales in north central Egypt (see chapter 2 and fig. 2.1). The second body type seen in dorudontines is shorter (as little at 4 m or 13 ft) and more dolphin-like. The hind limbs of basilosaurids are tiny and were not used in locomotion; swimming was likely tail-based as in extant cetaceans.

CROWN CETACEA (NEOCETI)

Neoceti is the taxonomic group comprising the two extant clades of whales: Odontoceti and Mysticeti (fig. 4.2). Crown cetaceans differ from stem whales in having a **telescoped skull**, in which the nostrils have moved to the top of the head, where they form the **blowhole**. Air and water vapor released from the blowholes, called the **blow**, form shapes that are species-specific and useful in field identification. Recently, scientists found that DNA samples can be obtained from the blow, providing a noninvasive method of genetic identification.

Odontoceti

As the name suggests, odontocetes possess teeth, although these differ in both number and degree of development. In odontocetes, telescoping of the skull involves the premaxilla and maxilla bones, which extend posteriorly and laterally to override the frontal bones (which form the skull roof) and crowd the parietals laterally (fig. 4.4). Relationships among stem and crown groups of odontocetes are shown in fig. 4.5.

STEM ODONTOCETES

The oldest described odontocete is *Simocetus rayi* (fig. 4.6) from the early Oligocene (32 Ma) of the North Pacific, although there are reports of undescribed late Eocene odontocetes. The downwardly arched upper jaw of *Simocetus*, with its toothless tip, may have been used in grubbing through

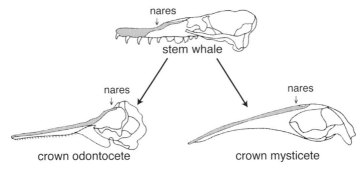

nares
↓

stem whale

nares
↓

nares
↓

crown odontocete

crown mysticete

Figure 4.4. Telescoping of the skull in cetaceans (modified from Berta et al. 2006). Note the posterior position of the nares and arrangement of premaxilla (shaded) and maxilla; the latter is directly below the premaxilla.

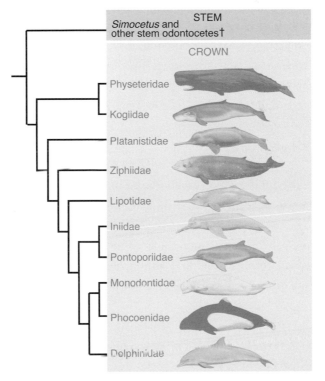

Figure 4.5. Evolutionary relationships of major lineages of odontocetes.

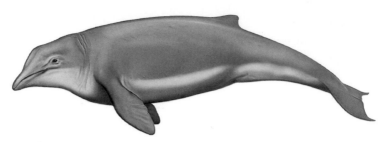

Figure 4.6. Stem odontocete (*Simocetus rayi*) life restoration (painted by Carl Buell).

the sediments. Another purported early odontocete not closely related to *Simocetus* is *Agorophius* from the late Oligocene (24 Ma) of South Carolina. However, the **holotype**, the diagnostic specimen that is the name bearer for the taxon, is missing and some scientists have suggested this may not be a stem odontocete as once believed. Better-known stem odontocetes include *Archaeodelphis* and *Xenorophius* from the late Oligocene. Both taxa have specialized facial structures and the apparent lack of development of nasofacial muscles in *Archaeodelphis* may indicate poor echolocation abilities.

CROWN ODONTOCETES

Crown odontocetes include 10 families: Ziphiidae, Physeteridae, Kogiidae, Platanistidae, Pontoporiidae, Iniidae, Lipotidae, Delphinidae, Phocoenidae, and Monodontidae. There are at least 73 species within these 10 families. Relationships among major lineages of crown odontocetes are fairly well resolved (fig. 4.5) although species-level relationships within some groups are still debated (for example, Delphinidae).

ZIPHIIDAE (BEAKED WHALES)   Beaked whales are so named for their snout, which is drawn out into a beak. They are a diverse group, with at least six genera and 21 extant species described. Ziphiids are among the least-known whales and they inhabit deep ocean basins. They are moderately large in size, ranging from 4 to 13 m (13–43 ft) in length, and weighing from 1 to 14 tons. All ziphiids possess a pair of grooves on the throat that aid in suction-feeding squid and fish. Ziphiid males exhibit an evolutionary trend toward the loss of all teeth in the upper jaw and most in the lower jaw, with the exception of one or two pairs of teeth in the front of the jaw (incisors), which are often very enlarged as tusks. They are one of two cetacean species with tusks, and scientists have long wondered why since their diet is primarily squid. Research has shown that the male's teeth are actually a secondary sexual characteristic, similar to the antlers of male deer. Each species has characteristically and uniquely shaped teeth; because the different species are otherwise quite similar in appearance, the females cue on the teeth to select the proper males with whom to mate.

PHYSETERIDAE (SPERM WHALE)   Sperm whales are perhaps best known from Herman Melville's classic American whaling novel *Moby Dick*. Sperm whales have an ancient and diverse fossil record, although only a single species (*Physeter macrocephalus*) survives (fig. 4.7). The term sperm whale derives from the erroneous belief of those who named the whale that it carried sperm in its head. Instead, this milky white tissue filled with waxy fluid found in the head is **spermaceti**, once considered a valuable resource by whalers for candle making and burning in lanterns. Various functions have been attributed to the spermaceti organ. At one time, it was hypothesized to regulate buoyancy and aid in deep diving. Other scientists have suggested that it evolved as a battering ram in aggressive encounters between males. Currently, the best-supported hypothesis suggests that the spermaceti organ is involved in echolocation. Sperm whales are the largest of the toothed whales, attaining a length of as much as 19 m (62 ft) and weighing 70 tons. A sperm whale has a large head that can be up to one-third of the animal's length. They are also the longest- and deepest-diving vertebrates known, 138 min and 3,000 m (more than 1.8 mi) on a single breath!

Discovery of the largest fossil sperm whale (*Lyviatan melvillei*) in 12 Ma rocks from Peru with 36 cm (1 ft long!) teeth represents one of the

Figure 4.7. Cuvier's beaked whale (*Ziphius cavirostris*), Ziphiidae; sperm whale (*Physeter macrocephalus*), Physeteridae; and pygmy sperm whale (*Kogia breviceps*), Kogiidae. Painted by Carl Buell.

largest-known predators, with an estimated body size of 13.5–17.5 m (43–59 ft; see fig. 4.8). *Lyviatan* is suggested to have fed on baleen whales, differing from the living sperm whale, a proficient squid feeder.

KOGIIDAE (PYGMY AND DWARF SPERM WHALES) Pygmy sperm whales (*Kogia breviceps*) and dwarf sperm whales *(Kogia simus)* are close relatives of the sperm whale family, Physeteridae. Male pygmy sperm whales attain a length of only 4 m (13 ft) and the dwarf pygmy sperm whale is even smaller at less than 3 m (10 ft) (fig. 4.7). The presence of a "false gill" gives them a superficial resemblance to sharks. They are widely distributed in tropical and temperate regions. The oldest kogiids are from the late Miocene of South America.

Figure 4.8. An artist's impression of a raptorial sperm whale (*Lyviatan melvillei*) attacking a baleen whale (Lambert et al. 2010, courtesy of C. Letenneur).

RIVER DOLPHINS Living river dolphins include four species in four lineages (families): Platanistidae, Lipotidae, Iniidae, and Pontoporiidae, which have invaded estuarine and freshwater habitats. Although river dolphins are superficially similar (for example, long rostra, flexible necks, and reduced eyes), they have independently adopted a freshwater or estuarine habit. They are characterized by having numerous teeth; typically, those in the anterior portion of the jaw are narrow and pointed for seizing prey and those in the rear are short and flattened for crushing heavily scaled fish.

PLATANISTIDAE (ASIATIC RIVER DOLPHIN OR SUSU)   The extant Asiatic river dolphin (*Platanista gangetica*) includes two subspecies: the Ganges river dolphin (*Platanista gangetica gangetica*) and the Indus river dolphin (*Platanista gangetica indus*) which constitute the Platanistidae. This is the only river dolphin to exclusively occupy freshwater. Females are larger than males, ranging from 2.4 to 2.6 m (8.0–8.5 ft), compared to males, which range from 2.0 to 2.2 m (6.5–7.2 ft).

The skull is distinctive in the development of a bony facial crest formed of maxillary bone, which encloses the melon, perhaps used like a megaphone to direct sound. The species is effectively blind, although it may still be able to detect light intensity and direction. They swim on their sides using broad, paddle-shaped flippers. There is no fossil record of the crown group, but stem platanistids are known from the Miocene (6–16 Ma) of the North Atlantic and Paratethys.

PONTOPORIIDAE (FRANCISCANA OR LA PLATA DOLPHIN)   The sole living member of this lineage, the small, long-beaked franciscana (*Pontoporia blainvillei*) lives in temperate coastal Atlantic waters of southeastern South America, ranging from Brazil to Argentina. It is the only river dolphin that actually lives in the ocean and estuarine waters rather than freshwater. It shares with other river dolphins the presence of a long, slender beak with numerous teeth (fig. 4.9). Distant relatives of *Pontoporia* indicate a past broader distribution that included Southern California during the Pliocene.

INIIDAE (BOUTO OR PINK RIVER DOLPHIN)   The bouto (*Inia geoffrensis*) is a freshwater species found only in the Amazon and Orinoco River drainages of Venezuela, Peru, Brazil, Bolivia, and Colombia. The name comes from the sound of its blow. The unique pink coloration is not well understood but may be due to temperature and mineral content (that is, iron) of the water. It is the largest of the river dolphins, at 1.8–2.5 m

Figure 4.9. Ganges river dolphin (*Platanista gangetica*), Platanistidae; franciscana (*Pontoporia blainvillei*), Pontoporiidae; bouto (*Inia geoffrensis*), Iniidae; and Chinese river dolphin (*Lipotes vexillifer*), Lipotidae. Painted by Carl Buell.

(6–8 ft) in length, weighing 85–150 kg (187–330 lb; see fig. 4.9). Three subspecies have been described, each occupying different river systems (Orinoco, Amazon, and Madeira). The fossil record of iniids goes back to the late Miocene of South America

LIPOTIDAE (BAIJI OR CHINESE RIVER DOLPHIN)   The baiji or Chinese river dolphin (*Lipotes vexillifer*) occupied the Yangtze River, China. They are characterized by a long, upturned beak; a low, triangular dorsal fin; broad, rounded flippers; and very small eyes (fig. 4.9). The Baiji population declined drastically in recent decades as China industrialized and made heavy use of the river for fishing, transportation, and hydroelectricity. Efforts were made to conserve the species, but a late 2006 expedition failed to find any baiji in the river. The baiji is likely extinct, which would make it the first aquatic mammal species to become extinct since the demise of the Japanese Sea Lion and the Caribbean Monk Seal in the 1950s. They are unknown from the fossil record.

DELPHINIDAE (DOLPHINS)   Dolphins are the most diverse cetacean family and include 17 genera and 36 extant species of dolphins, killer whales, and pilot whales. There is much variation in body size and proportion, skull shape, and teeth. Most delphinids are small to medium-sized, ranging from 1.5 to 4.5 m (5 to nearly 15 ft) in length (fig. 4.10). The giant among them, the killer whale, reaches 9.5 m or nearly 32 ft in length.

Archaic dolphins known from the Miocene in both the Atlantic and Pacific are included in one of three families: Kentriodontidae, Albeirodontidae, and the Eurhinodelphinidae. Eurhinodelphids, with their characteristically very long beaks and toothless premaxilla, which may have been used to hit or stab prey, were moderately diverse and widespread.

MONODONTIDAE (BELUGA AND NARWHAL)   Monodontids include two extant species: the narwhal (*Monodon monoceros*) and the beluga or white whale (*Delphinapterus leucas*). Although at the present time both the narwhal and beluga have a circumpolar distribution in the Arctic, during the Miocene monodontids occupied temperate waters as far south as Baja California, Mexico.

The most distinctive feature of the beluga is the pure white coloration of adults, allowing them to readily blend into their Arctic

Figure 4.10. Bottlenose dolphin (*Tursiops truncatus*), Delphinidae; and narwhal (*Monodon monoceros*). Painted by Carl Buell.

environment. Belugas lack a dorsal fin and they are unique among toothed whales in having unfused neck vertebrae, allowing flexibility of the head and neck. Unlike most other toothed whales, most belugas migrate from icy Subarctic and Arctic waters to winter grounds in warm water bays and estuaries. Likely related to their annual migration is another unique feature of belugas distinguishing them from other cetaceans, an annual molt (see also chapter 3).

The narwhal (*Monodon monoceros*) is readily distinguished by the presence of a long (2.7 m or up to 9 ft), spiraled left incisor tusk in males and occasionally in females (fig. 4.10). Historians believe that the narwhal tusk may have been the inspiration for the unicorn myth. Many hypotheses for the narwhal tusk have been proposed, including breaking ice, spearing fish, attracting a mate, functioning in male–male competition bouts, or defending against a predator. Recent work revealed nerve endings in the tusk and suggests that the tusk might also serve as a temperature, pressure, or water quality sensor. In a novel use of oceanographic instrumentation, narwhals outfitted with thermometers have provided climate data from Baffin Bay, an area previously without winter temperature data.

An extinct relative of monodontids (*Odobenocetops*) lived in Peru during the early Pliocene (fig. 4.11). The presence of tusks and a presumed mollusk-eating, suction-feeding habit are convergences *Odobenocetops* shares with the walrus.

PHOCOENIDAE (PORPOISES)   Porpoises include six small extant species. They are closely related to dolphins (delphinids). The most obvious visible difference between the two groups is that porpoises have shorter beaks and flattened, spade-shaped teeth distinct from the conical teeth of dolphins (fig. 4.12). Porpoises are among the smallest cetaceans, ranging from the vaquita at 1.5 m (4.9 ft) in length and weighing up to 55 kg (120 lb) to Dall's porpoise at up to 2.3 m (7.5 ft) in length and weighing up to 200 kg (440 lb). Another feature of porpoises is **paedomorphosis,** a process that stems from evolutionary changes in rates and timing of development. In porpoises, paedomorphosis results in the retention of juvenile characters in adults in their skeletal morphology. It has been hypothesized to be associated with small size and the rapid life history patterns found in porpoises.

Mysticeti

Mysticeti, or moustached whales, are so named for their feeding apparatus, baleen, which hangs in racks suspended from the roof of the mouth and is used as a filter to strain fish and zooplankton. Although mysticetes are born with teeth, these are rudimentary and resorbed before birth. At the same time as the teeth degenerate, baleen begins to develop. Telescoping of the skull in mysticetes differs from that of odontocetes in having the maxilla extend posteriorly underneath the frontal (fig. 4.4). Relationships among stem and crown groups of mysticetes are shown in fig. 4.13.

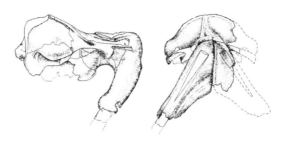

Figure 4.11. *Odobenocetops peruvianus* skull and life reconstruction (Muizon 1993 [top], and bottom courtesy of M. Parrish).

Figure 4.12. Dall's porpoise (*Phocoenoides dalli*), Phocoenidae; painted by Carl Buell.

STEM MYSTICETES

Both toothed and baleen-bearing taxa are stem mysticetes (fig. 4.13). Neither group is monophyletic. Toothed mysticetes first evolved in the late Eocene or earliest Oligocene, diversified in the late Oligocene, and appear to have gone extinct before the Miocene. Toothed mysticetes are grouped into three families: Llanocetidae and Mammalodontidae from the Southern Ocean and Aetiocetidae from the North Pacific (fig. 4.14). The oldest reported mysticete is *Llanocetus denticrenatus*, a large animal (with a skull length of 2 m or more than 6.5 ft) with multi-cusped teeth, reported from the late Eocene or early Oligocene (34 Ma) of Antarctica. The exact role of the widely spaced teeth in feeding is uncertain, but it is possible that spaces between the teeth were occupied by baleen. Other stem mysticetes, *Mammalodon* and *Janjucetus*, show some skull specializations not seen in later mysticetes, such as very small, short-faced skulls, large eyes, and well-developed, robust occluding teeth. Aetiocetids are the most diverse toothed mysticetes and include seven named species grouped into three to four genera. Overall, aetiocetids were small-bodied cetaceans with skull lengths of about 60–70 cm (2–2.3 ft) and an estimated total body length of 2–3 m (6.5 to nearly 10 ft). Unlike *Mammalodon* and *Janjucetus*, aetiocetids had a relatively long rostrum. In addition to teeth, aetiocetids appear to have also possessed baleen (fig. 4.14).

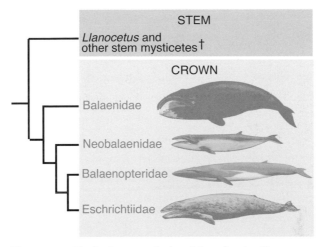

STEM

Llanocetus and
other stem mysticetes†

CROWN

Balaenidae

Neobalaenidae

Balaenopteridae

Eschrichtiidae

Figure 4.13. Evolutionary relationships of major lineages of mysticetes.

Figure 4.14. Reconstruction of *Aetiocetus weltoni*, showing hypothesized simultaneous occurrence of baleen and teeth (painted by Carl Buell).

Baleen-bearing stem mysticetes include several extinct lineages. The earliest known baleen-bearing mysticete (*Eomysticetus*) is known from the late Oligocene of South Carolina. These earliest baleen-bearing mysticetes had long, narrow skulls. Other probable eomysticetids include *Mauicetus* from the late Oligocene of New Zealand and undescribed specimens from the North Pacific. "Cetotheriidae" is a large, diverse, nonmonophyletic (indicated by quotes) assemblage. "Cetotheres" comprise the greatest taxonomic and morphologic diversity among fossil mysticetes, with over forty-five described species divided among more than 30 genera. "Cetotheres" had a worldwide distribution from the late Oligocene to the late Pliocene in North America, South America, Europe, Japan, Australia, and New Zealand. Several recent phylogenetic studies have recognized two alternate topologies; either "cetotheres" lie outside crown mysticetes or they are variously positioned within crown mysticetes.

CROWN MYSTICETES

Crown mysticetes include four clades: Balaenidae (bowhead and right whales), Balaenopteroidea (fin whales or rorquals), Eschrichtiidae (gray whales), and Neobalaenidae (pygmy right whales). There is controversy about whether neobalaenids and balaenids are united in a clade, which is supported by morphology, or if neobalaenids are sister to all other mysticetes, based on molecular data. Further, morphology also supports an alliance between eschrichtiids and balaenopterids (Balaenopteroidea), while molecular data nests eschrichtiids with balaenopterids, thus rendering the latter nonmonophyletic (fig. 4.13).

BALAENIDAE (BOWHEAD AND RIGHT WHALES) Balaenids include right whales and bowheads. A single species of bowhead (*Balaena mysticetus*) is found throughout the Arctic and Subarctic. They are large and robust with very thick blubber layers (as much as 50 cm or 1.5 ft thick; see fig. 4.15). Bowheads were extensively hunted until the early twentieth century.

The name *right whale* came from hunters, with reference to the fact

Figure 4.15. Bowhead (*Balaena mysticetus*), Balaenidae. Painted by Carl Buell.

that these whales inhabit coastal waters, are slow swimmers, and float when dead, making them the "right" whale to hunt. Three species of right whale are currently recognized; two species, North Atlantic right whales (*Eubalaena glacialis*) and North Pacific right whales (*Eubalaena japonica*), occupy the Northern Hemisphere. The third species, the southern right whales (*Eubalaena australis*), are circumpolar in the Southern Hemisphere. The North Pacific and southern right whales are more closely related to one another than either is to North Atlantic right whales.

The most conspicuous external characteristic of right whales is the presence of **callosities**, raised patches of roughened skin, found on the head. These callosities are inhabited by **whale lice** and barnacles. Whale lice are actually not related to the lice of terrestrial animals but are specialized cyamids, crustaceans present on the skin of most species of whale. The callosity patterns are unique to individuals and thus useful as a natural marker. Since callosities stay with an animal throughout its life, they have been used in the photo identification of individual whales (see chapter 1). Balaenids are also characterized by tall mouths with highly arched upper jaws that accommodate their exceptionally long (up to 4 m or 13 ft) baleen plates.

NEOBALAENIDAE (PYGMY RIGHT WHALE) Neobalaenids are represented by a single living species, the pygmy right whale (*Caperea marginata*), which lives in temperate and Subantarctic regions of the Southern

Figure 4.16. Pygmy right whale (*Caperea marginata*), Neobalaenidae. Painted by Carl Buell.

Hemisphere. As their name suggests, pygmy right whales are a small baleen whale, approximately 4 m (13 ft) in length (fig. 4.16). The skull and skeleton of *Caperea* are unlike those of other cetaceans, possessing more ribs and fewer vertebrae than other whales and having a skull with a short, wide rostrum and a forward projecting occipital shield.

BALAENOPTERIDAE (RORQUALS) Balaenopterids, commonly called the rorquals (from the Norwegian word *røyrkval*, meaning furrow whale, with reference to their throat grooves), are the most diverse and abundant baleen whales (fig. 4.17). Balaenopterids include two genera: *Balaenoptera*, with seven to eight species, and *Megaptera*, represented by a single living species. They range in size from the small 9-m (approximately 30-ft) common minke whale (*Balaenoptera acutorostrata*) to the giant blue whale (*Balaenoptera musculus*). The blue whale has the distinction of being the largest mammal ever to have lived, reaching 33 m (or more than 100 ft) in length and weighing over 160 tons. Their circulatory system pumps 10 tons of blood through the body, using a one-ton heart (that is, the size of a small car).

The oldest named crown balaenopterid is "*Megaptera*" *miocaena* from the late Miocene of California. Phylogenetic study suggests that this taxon is likely not a close relative of the humpback whale (*Megaptera novaeangliae*). Humpback whales have a distinct body form with a knobby head, long pectoral flippers, and small dorsal fin. They typically undertake long annual migrations between feeding and breeding grounds. The varying fluke patterns provide unique visual identity and they are

Figure 4.17. Fin whale (*Balaenoptera physalus*), Balaenopteridae, and gray whale (*Eschrichtius robustus*), Eschrichtiidae. Painted by Carl Buell.

a common species for photo identification (see chapter 1). They are perhaps best known for their powerful, evocative songs (see later discussion).

Rapid diversification of balaenopterids occurred during the late Miocene and Pliocene. Relationships among extant species based on both molecular and morphologic data suggest an earlier divergence of common and Antarctic minke whales (*B. acutorostrata* and *B. bonaerensis*) followed by divergence of a clade that includes the sei whale (*B. borealis*), blue whale (*B. musculus*), and Eden's whale (*B. edeni*), which is sister to the humpback and fin whales (*Megaptera* and *Balaenoptera physalus*).

ESCHRICHTIIDAE (GRAY WHALE)   Eschrichtiids are represented by a single living species, the gray whale (*Eschrichtius robustus*). Gray whales are characterized by lack of a dorsal fin, which is replaced by a series of dorsal humps along the back and two to four throat grooves (fig. 4.17). The gray whale has a fossil record that goes back to the Pleistocene (200,000–300,000 years) and the genus *Eschrichtius* extends back to the late Pliocene (1.5–3.5 Ma). Once found throughout the Northern Hemisphere, the gray whale is now confined to the North Pacific Ocean and adjacent waters of the Arctic Ocean. There are two Pacific populations of gray whale; the eastern population occurs in the North Pacific and Arctic Oceans and the western population occurs in the western North

Pacific off the Asian coast. A third population of gray whales also once occupied the North Atlantic. The last few gray whales in the North Atlantic are believed to have been hunted to extinction by whalers near the end of the 17th century. Recent reports of a gray whale sighted in the Mediterranean for the first time in nearly 300 years suggest one of two explanations. The most likely possibility is that a wayward whale traveled a circuitous route from the Pacific across the Canadian Arctic via a corridor facilitated by melting ice. A less likely possibility is that there has been a relict population of gray whales in the Atlantic that has not been observed previously.

EVOLUTIONARY TRENDS

Early Eocene whales had four limbs, heterodont teeth, and shared cranial, vertebral, and forelimb characters with crown cetaceans. By the middle Eocene, dorudontine whales approached crown cetaceans in body form, having a tail fluke, a strongly shortened neck, nearly absent hind limbs, and a diversity of homodont tooth types, including the loss of all dentition (like some extant beaked whales). Their method of locomotion changed from a quadrupedal terrestrial lifestyle to semiaquatic, aquatic, and marine living, as revealed by study of the postcranial skeleton. Study of the semicircular canal system housed in the middle ear reveals adaptations for agile swimming early in whale evolution.

The radiation of Neoceti is probably related to changes in oceanic circulation that resulted in increased productivity of the oceans (see also chapter 2). The evolution of odontocetes is closely tied to the origin of **echolocation**, the ability to produce high-frequency sound and receive its reflected echos. The oldest odontocetes show evidence in their facial structures of having been able to echolocate. It has been suggested by paleontologists Lindberg and Pyenson that echolocation in early odontocetes was initially an adaptation for feeding at night on vertically migrating cephalopods, especially nautiloids, marine mollusks with external shells. Evidence for this coevolution between predator (odontocetes) and

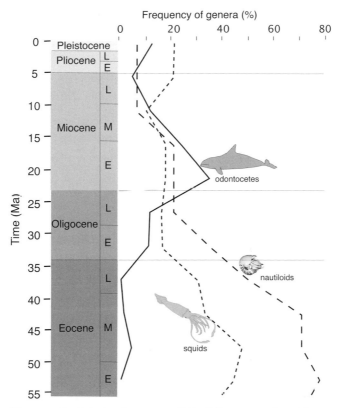

Figure 4.18. Relative frequency (%) of Cetacea and squid during the Tertiary; coevolution of predator and prey (modified from Lindberg and Pyenson 2007).

prey (nautiloids) includes the fact that gas-filled nautiloid shells produce stronger echos than soft-bodied cephalopods like squid. These Oligocene nautiloids would have been easily detectable prey for early echolocating odontocetes, and this vulnerability may have been responsible for their near demise (fig. 4.18). Subsequent modification of this echolocation system, as documented in the fossil record, was driven by odontocetes hunting squid and other prey deeper in the water column.

While the earliest mysticetes had teeth that were used in feeding, these teeth became reduced in size and were eventually lost. Genetic studies have shown that the dental genes responsible for tooth formation were inactivated in the common ancestor of baleen whales. Other major evolutionary trends within mysticetes include the development of large body size, laterally bowed lower jaws, moveable skull joints, and large mouths with racks of baleen used in filter feeding. Mysticetes employ low-frequency hearing in communication, which is likely a modification of the ear of early whales and an adaptation for long-distance communication. Body-size evolution suggests that some mysticetes were of relatively large body size (5–12 m or 16 to nearly 40 ft) ancestrally (for example, *Llanocetus*), with independent evolution of smaller body sizes in other stem mysticetes, such as mammalodontids and aetiocetids and several later clades. The small body size and other character reversals (that is, relatively large orbits, low cranium) of archaic mysticetes have been used to indicate the retention of juvenile traits, or paedomorphism.

While key evolutionary innovations, such as sonar and baleen, have been suggested as triggers for toothed whale and baleen whale diversification, respectively, the tempo or rate of evolution has only recently been critically examined. Results of a study that examined the causes of whale diversification revealed that species diversity and variation in body size were established early in whale evolution (25 Ma). Large whales, small whales, and intermediate-sized whales all appeared early in the history of whales, with large whales eating mostly zooplankton, small whales eating fish, and intermediate-sized whales eating squid. For the most part, these same body size niches exist today. Killer whales are an exception, having become larger over the last 10 million years, but their food source is atypical, since some eat mammals, including other whales.

While it is true that whales show a range of body sizes, when compared to other diving vertebrates such as fish, whales are considerably larger in size. The reason whales are large may have a lot to do with their metabolism. Whales are air-breathing **endotherms** that maintain a constant body temperature, opposed to **ectotherms** such as fish, whose temperature is dependent on the environment. Evidence also suggests that this difference in size and metabolism between endotherms and

FILTER FEEDERS IN THE PAST

Filter feeding did not begin with baleen whales. The filter-feeding niche exploited today by baleen whales was likely occupied in the past (170–65 Ma) by now-extinct giant bony fish and plankton-feeding sharks and rays using gill rakers as filters instead of baleen. The demise of many of these large-bodied marine fish at the end of the Cretaceous likely created an ecological opportunity for the radiation of suspension-feeding rays and sharks in the mid-Eocene and filter-feeding whales in the late Eocene and early Oligocene (fig. 4.19). Their diversification was linked to the diversity of their prey, diatoms, an important marine primary producer. An increase in diatom diversity has been linked to an increase in the availability of silica and other nutrients in the Southern Ocean and coastal upwelling zones around the world (see also chapter 2).

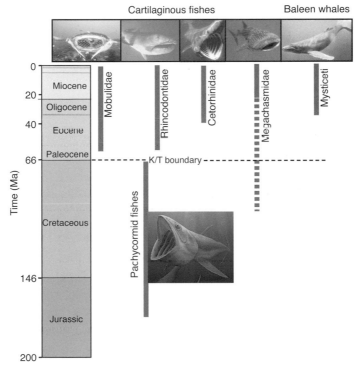

Figure 4.19. Ecologic replacement of various fish lineages with baleen whales (modified from Friedman et al. 2010).

ectotherms is associated with swimming speed. Larger divers such as whales were found to swim faster than fish. An extrapolation that requires further testing suggests that selective pressures may have led whales to exploit deeper waters and search for food more efficiently than other similar-sized aquatic vertebrates.

STRUCTURAL AND FUNCTIONAL INNOVATIONS AND ADAPTATIONS
Skin with Thick Blubber

The outer layer of skin, the **epidermis**, in cetaceans is smooth and rubbery to the touch. The middle layer of skin, the **dermis**, of cetaceans is distinguished from that of all other marine mammals in lacking hair follicles. The inner layer of skin, the hypodermis or blubber, is loose connective tissue composed of fat cells and collagen fibers. Blubber thickness varies by age and individual as well as seasonally. Blubber provides insulation and fat (energy) storage as well as smoothing out the body shape.

Color Patterns

The most widespread color pattern in cetaceans is countershading, a form of camouflage by which predators do not see a contrast between the animal and the environment (see also chapter 3). Some species, like the killer whale (see fig. 1.4), have striped patterns with sharply colored black and white areas on the head, side, belly, and flukes. Such distinctive markings function in individual recognition, including the visual coordination of animals when traveling in groups. The uniform white color of belugas functions in concealing them from potential predators on the Arctic ice.

Bending in Whales

Under the dermis layer of the skin is a subdermal connective tissue sheath made of collagen (protein) fibers that serves to restrain the vertebrae and provides an anchor for muscles that flex and extend the tail. Most whales have two or more neck vertebrae that are fused, which limits mobility and stabilizes the head. Several odontocete lineages, including pontoporiids, platanistids, iniids, and balaenopterid mysticetes possess unfused neck vertebrae, which allows neck flexibility and likely aids in foraging in narrow places.

Archaic whales possessed tail vertebrae and a pelvic girdle for hind limb attachment, elements that became progressively reduced and eventually lost during their evolution. The beginning of the fluke is marked by the presence of chevron bones, ventral processes on the posterior (lumbar) vertebrae. Chevron bones function to protect and support the tail.

How to Make a Flipper and a Fluke?

Although modern whales normally lack hind limbs, their embryos still show the beginnings of hind limbs. This is due to disruption of the normal expression of **Hox genes,** which determine the basic body plan (head to tail) organization of an organism. Very rarely, an anomalous individual with hind limbs is born with external vestiges of hind limbs, the result of a mutation that allowed a developmental pathway for limb development to be reactivated.

FLUKES AND SWIMMING

The tail or **fluke** of cetaceans is mostly composed of tough, fibrous connective tissue. Its shape differs in response to varying hydrodynamic parameters although most are designed to provide high lift. The trailing edges of most are tapered but some are straight (sperm whale), conspicuously curved (humpback), or even slightly concave (narwhal). The underside of the flukes as well as the dorsal fins of some whales develop distinctive scars and markings with time, which enables identification of individuals in the field.

Cetaceans, as well as sirenians (see next chapter), swim by

dorsoventral undulations of the tail. There is no recovery phase and high levels of thrust are produced on both the upstroke and downstroke (fig. 4.22). Whales, especially dolphins, conserve energy by gliding rather than swimming at depth. This switching of modes of locomotion, from passive gliding to active swimming, is similar to terrestrial mammals switching between gaits (a horse trot versus a gallop).

EXTRA FINGERS AND DIGITS
IN THE WHALE FORELIMB

Some whales (for example, pilot whales, humpbacks) have increased the structural support of the flipper and hence its hydrodynamic efficiency by adding extra finger bones, a process called **hyperphalangy** (fig. 4.20). Recall that pinnipeds also have lengthened the fingers by the addition of pieces of cartilage to the tips of existing finger bones.

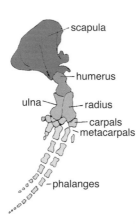

Flipper shape reflects locomotor requirements. The narrow, elongate flippers of humpback whales facilitate fast swimming at the cost of reduced maneuverability, while the broad flippers of bowhead and right whales aid in slow turns. Supporting evidence was provided by comparative study of whale and dolphin flippers, which revealed that the bottlenose dolphin's triangular flipper was the most efficient hydrodynamically and capable of generating lift like the modern Delta wing aircraft.

Humpback whales have the longest flippers of any whale, with a length that is as much as one-third of total body length. These

Figure 4.20. Forelimb of pilot whale (*Globicephala* sp.) showing hyperphalangy (see Berta et al. 2006 for original data source).

SAVING MORE ENERGY: BOW OR
WAVE RIDING AND PORPOISING

To make forward progress, a swimming marine mammal must over-come the drag of water. To help minimize drag on the body when swimming, cetaceans have developed a number of behavioral strate-gies. Small cetaceans, such as dolphins, moving at high speed near the water's surface leap above the water and glide airborne for several

flippers are distinguished by the presence of large irregular bumps or tubercles on the leading edge of the flipper (fig. 4.21). The number and posi-tion of tubercles on the flipper function similarly to the alula in birds (small projection on the anterior edge "thumb" of wing) to maintain lift and reduce drag in the water. Humpback whale tubercles, with their drag reduction efficiency, have been adapted to the design of the leading edge of wind tur-bine blades, an example of a nature inspired design termed **biomimicry**. These long flippers are also very maneuverable and are "waved," especially during bubble cloud feeding and social displays.

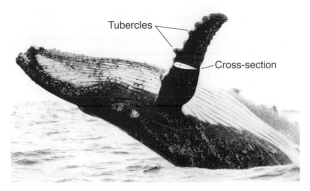

Figure 4.21. Flipper of humpback whale showing tubercles and cross-section illustrating hydrodynamic design (Fish and Battle 1995, courtesy of F. Vallejo).

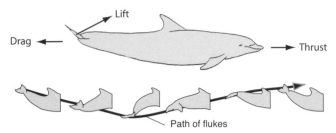

Figure 4.22. Cetacean locomotion tracings of flukes of bottlenose dolphin (see Berta et al. 2006 for original data source).

body lengths (termed **porpoising**). Another strategy is **wave or bow riding**, which is best described as surfing the bow or stern wave of the boat.

FINS, FLUKES, AND THERMOREGULATION

Most whales possess a prominent dorsal fin on their backs. The **dorsal fin** is similar to the fluke in its connective tissue composition. The dorsal fin of whales, like that of fish, serves as a keel and acts to prevent rolling during swimming. Dorsal fins vary in size and shape. For example, the dorsal fin of the male killer whale is quite large, as much as 1.8 m (6 ft!) high. The droopy or collapsed dorsal fin of some orcas in captivity may be the result of one or more factors, including less water pressure on the collagen in the fin in tanks than that in the ocean as well as the animal's genetics. Flippers, as discussed previously (see chapter 3), as well as flukes and dorsal fins, serve as thermoregulators, acting as radiators to regulate heat loss.

Big Brains, Encephalization Quotients, and Intelligence

A feature of some whale brains, similar in structure to the human brain, is their large size, especially the cerebrum. Brain size is usually expressed with respect to body size, the **encephalization quotient**

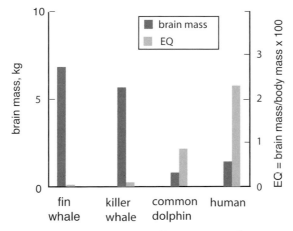

Figure 4.23. A comparison of brain mass and encephalization quotients (EQs) of several cetaceans compared to humans (based on data in Berta et al. 2006, table 7.1).

or EQ, and is a useful measure of brain evolution. Most odontocetes have relatively large brains and a high EQ. In fact, the dolphin brain is similar in size to a gorilla's brain. Only the human brain is proportionally larger. Although the question of exactly how brain size relates to intelligence is controversial, the complex cognitive abilities of dolphins are well known. EQ has been correlated with dietary strategy, social group structure, and life history patterns. The high EQs of odontocetes (with values greater than 1; see fig. 4.23) are partly explained by their complex social structure and behavior.

Senses

VISION AND SPY HOPPING

Some early whales, such as the stem mysticete *Janjucetus*, had large eyes, which suggests that vision was likely the most well-developed sense in the early evolution of whales. Cetacean eyes differ from those of pinnipeds (except the walrus) in their location on the side of the head and in being markedly flattened anteriorly. This corrects for the nearsightedness that would result from refracting (bending) of light rays as they go from water to air. It seems unlikely that whales have true color vision although they may possess a limited capacity for color discrimination.

The eye of the Ganges river dolphin is very small and lacks a lens, adaptations for living in murky, muddy water. With its poor eyesight, it is sometimes referred to as the blind dolphin.

Spy hopping is a fairly common behavior in which whales rise partially out of the water. Spy hopping is hypothesized to permit a view of prey or other whales if the rostrum and whale's eyes are above the surface of the water.

HEARING AND SOUND PRODUCTION

Although cetaceans have developed an acute sense of hearing, they possess very small external ear openings, suggesting that these openings may not be that important in conducting sound. In fact, the ear opening of baleen whales leads to a narrow ear canal that is completely filled with a waxy substance, and it is very small in toothed whales.

ECHOLOCATION

The ability to produce and receive high-frequency sounds for navigation and location of prey, known as echolocation, is developed in only two mammalian groups: some bats and toothed whales. Although different in their details, the echolocation abilities of both groups rely on the same changes to the same gene—prestin—a remarkable example of convergence at the molecular level.

Humans hear sounds in the range of 0.02–17 kHz and produce sounds in the larynx or voice box located in the neck. Odontocetes produce ultrasounds (that is, frequencies above 17 kHz) from the nasal region. The sounds are emitted through the "phonic lips"—paired,

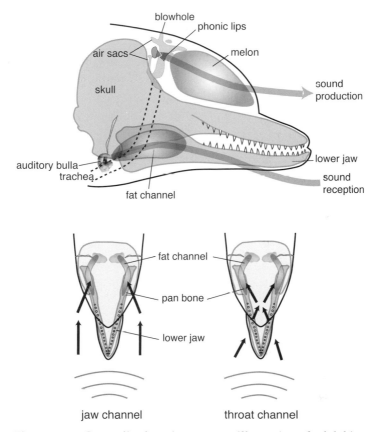

Figure 4.24. Generalized semitransparent illustration of a dolphin head showing sound production (nasal passages) and reception (jaw channel and throat channel; from Cranford et al. 1996, with modification by T. Cranford and C. Barroso).

fat-filled structures suspended within a complex array of muscles and air spaces near the blowhole. Sounds are focused by the melon, a fatty structure on top of the head, and returning echos pass through and under the lower jaw fat bodies before being transmitted to the middle and inner ear (fig. 4.24).

Three major classes of sounds are produced by odontocetes: (1) clicks or burst pulses, (2) whistles, and (3) pops or bangs (fig. 4.25).

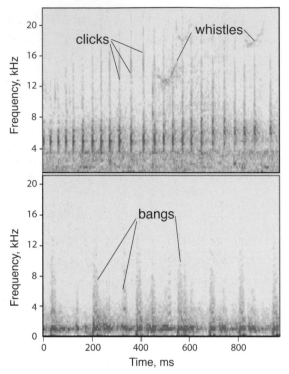

Figure 4.25. Comparison of clicks, whistles, and bangs on a frequency spectrum produced by a bottlenose dolphin (courtesy of V. Dudley).

Clicks are broadband frequency (most are in the range of 30–150 kHz), short-duration echolocation sounds that are principally used in navigation and foraging. Each click lasts less than a millisecond. The target discrimination capabilities of dolphins are impressive, unrivaled by any manmade sonar detection system. For example, results of experiments have shown that echolocating dolphins can discriminate targets that differ in thickness by only 0.27 mm (less than $\frac{1}{100}$ in) at a distance of 8 m (approximately 25 ft) 75 percent of the time. Sperm whales produce a variety of clicks for different contexts. For example, tightly spaced clicks, also known as **codas,** function in communication.

Codas appear to be produced only by sperm whales in social groups. Coda repertoires are shared between groups in a manner similar to killer whale dialects. Sperm whales also produce creaks, a rapid sequence of longer-duration clicks that sound to us like a continuous buzzing. They occur when a sperm whale is approaching a potential prey item. **Whistles** are narrow frequency band (0.5–40 kHz), long-duration sounds produced by odontocetes that function for individual recognition and to promote social cohesion. Whistles appear to be a learned behavior. **Pops** or bangs are low-frequency, loud impulse (10–25 decibels, a measure of sound pressure level, above maximum intensity recorded for dolphins) sounds observed in sperm whales and various dolphin species that have been suggested to stun or debilitate prey. A recent test of this prey debilitation hypothesis, however, did not provide supporting evidence, since no measurable changes were detected in the behavior of any fish species following exposure to the loud sounds.

SINGING WHALES

In contrast to toothed whales, baleen whales produce and hear low frequency sounds. Sounds are produced in the larynx or voicebox in the throat, as in other mammals. The actual mechanism and pathway of received sound transmission to the middle and inner ear are unknown. The songs of humpback whales are well known. Humpback whale songs are low-frequency (below 1.5 kHz), long, and complex and shared by all singing males on the breeding grounds. Songs of whales in various regions differ; for example, songs sung by humpback whales in the North Atlantic are different from those in the North Pacific. Each song has a hierarchical structure and is composed of repeated themes, phrases, subphrases, and units; the latter are typically a few seconds in duration. The structure of a humpback whale song changes slowly over time with deletion of old and insertion of new phrases. Study of male humpback calls suggests that, like songbirds, one male humpback may make another male change his tune. Also like songbirds, humpback whales appear to use songs to attract females, but more work is needed to determine the effect of male suitors and their different calls on females.

Study of several populations of blue whales has shown that over the last several decades they have been singing with deeper voices. Among several hypotheses proposed to explain this phenomenon, the leading hypothesis is that given an increase in their numbers since hunting ceased, they are able to sing at lower frequencies since their sounds don't need to travel as far to be heard. Since it takes more energy to sing deeply, it is possible that there is a selective advantage if singing attracts females, but this has not yet been demonstrated. Another possibility may be that it is the result of the changing dynamics of sound as it travels through water that has become warmer as the earth has heated up, but such small changes in temperature do not appear to match up with the larger shift in frequency.

BALANCE

Equilibrium, or balance, perhaps the least appreciated of our senses, is an essential component to the locomotor behavior of animals both on land and in the water. Study of the organ of balance, the semicircular canal system in the ear of whales, which is smaller than expected given their body size, suggests that whales have reduced the size of their balance organ to reduce their sensitivity. This has enabled them to accommodate large head movements and rapid changes in swim speed or direction.

Diet and Feeding Mostly with Teeth

Odontocetes are active hunters, pursuing mobile prey using echolocation (see later discussion). Some odontocetes, such as physeterids and ziphiids, primarily feed on squid, whereas others, including phocoenids and delphinids, are mostly fish eaters. As their name indicates, odontocetes, or toothed whales, possess simple, peg-shaped, homodont teeth that do not show the **heterodont** specialization seen in most other mammals. This is a function of their use in prey capture and transport to the stomach, where digestion occurs. There is considerable variation in tooth number both within and among species. Beaked whales show the most reduced dentitions among odontocetes, most species having lost all functional teeth with only a single pair of teeth occupying each side of the lower jaw. These teeth are likely used in male intraspecific fighting rather than in feeding (see section on Ziphiidae).

Filter Feeding: Gulpers, Skimmers, and Suction Feeders

Mysticetes feed primarily on shrimp-like crustaceans known as **krill**, copepods (fig. 4.26), and small fish. Their feeding apparatus consists of two rows of baleen plates that hang from the roof of the mouth. The plates grow continuously down from the gum but they are also continually worn down by the tongue and the prey items. **Baleen** is composed of keratin, the same material that makes up hair, claws, and fingernails of mammals. The number and length of the baleen plates on each side of the mouth range from about 155 (0.4– 0.5 m or 1.3–1.6 ft in length) in the gray whale to more than 350 in right whales (some exceeding 3 m, nearly 10 ft in length!). Water is expelled through the baleen by the tongue and the retained prey is swallowed.

Mysticetes have evolved three different types of feeding strategies: gulping, skimming, and suction feeding. Right and bowhead whales are skimmers with very long baleen. They feed by swimming slowly with their mouths open, constantly skimming very small prey, mostly copepods, from the water (fig. 4.26). Rorquals are fast swimmers that lunge at prey and gulp large amounts of water and prey, mostly krill and small fish. Expansion of the mouth and throat is facilitated by external throat grooves or pleats below the mouth and throat (fig. 4.26). With the mouth closed, water is forced out across the baleen so that food is trapped and

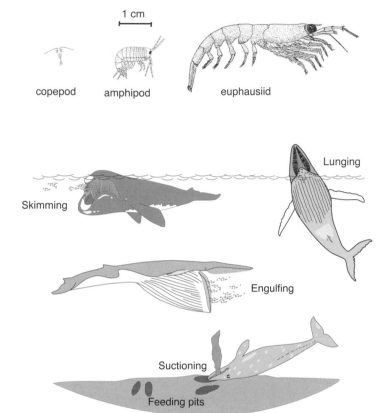

Figure 4.26. Lateral views of copepod, amphipod, and euphausiid (krill). Mysticete feeding types. Skim feeding by a right whale, engulfment feeding by a fin whale, lunge feeding by a humpback, and suction feeding by a gray whale (see Berta et al. 2006).

then swallowed. Gray whales, mainly bottom feeders, employ suction feeding, in which the animal rolls to one side and uses the tongue to draw water and sediment (fig. 4.26). The short, coarse baleen of gray whales is used to filter bottom-dwelling invertebrates, mostly amphipod and mysid crustaceans, from the sediments. Mysticetes and some other marine mammals undertake long-distance annual migrations between feeding and breeding areas, as discussed in chapter 1.

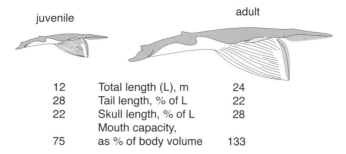

juvenile   adult

12	Total length (L), m	24
28	Tail length, % of L	22
22	Skull length, % of L	28
	Mouth capacity,	
75	as % of body volume	133

Figure 4.27. As a fin whale grows, its mouth does not scale linearly but takes up a larger percentage of its body size. Skull length increases from 22 percent to 28 percent of body length, whereas tail length decreases from 28 percent to 22 percent of body length; engulfment capacity rises from 75 percent to 133 percent of body mass (modified from Goldbogen 2010).

HOW MUCH DO THEY EAT?

Baleen whales, given their large size, have large daily food requirements (except when travelling or breeding). Rorquals, such as fin whales, can engulf a volume of water that is greater than their body mass. An adult blue whale can eat 40 million krill in one day. Another large mysticete, the fin whale, has been reported to engulf 70,000 liters (18,000 gal) of water in each gulp, containing 10 kg (22 lb) of krill. Over several hours of foraging, a whale can ingest more than a ton of krill, enough to sustain it for a day. The food consumed during these prodigious feeding bouts must sustain baleen whales for several months of winter fasting as well.

BIG MOUTHS AND MULTICHAMBERED STOMACHS

Some mysticetes (rorquals) have large jaws and mouths and small posterior body portions (dorsal fin to tail) relative to their body size. These proportional differences, known as **allometry**, indicate that the head grows at a faster rate than the tail, suggesting that this represents an adaptation in rorquals to increase their engulfment capacity as they mature (fig. 4.27).

All whales are characterized by a complex stomach with multiple divisions that resemble those of ruminant artiodactyls (cattle, sheep, goats). Found in the digestive system of sperm whales is a waxy substance called **ambergris**, which was used in making perfume during the whaling industry. Ambergris is hypothesized to form around the sharp beaks of squid, facilitating their passage through the digestive system.

Specialized Feeding Strategies

BUBBLE NET FEEDING

An unusual type of feeding strategy seen in humpback whales, termed **bubble net feeding**, involves the production of a series of bubbles in the form of clouds or nets to corral or trap prey. This type of strategy is seen often in tropical water with good visibility and seems to be used primarily on types of schooling fish such as herring. These bubbles may also act as camouflage, a visual barrier, to block a potential predator's view of the whale.

TOOL USE

In Shark Bay, Australia, some female Indo-Pacific bottlenose dolphins (*Tursiops aduncus*) have been observed collecting live sponges from the sea floor with their mouths and probing the sediments, effectively using sponges as foraging tools, a behavior termed **sponging** (fig. 4.28). It has been confirmed by behavioral and DNA studies that this behavior is passed down from mother to offspring, particularly daughters. This cultural acquisition of tool use suggests a degree of social learning in bottlenose dolphins that is similar to that observed in primate communities. Other foraging tactics of Shark Bay bottlenose dolphins include **kerplunking**, whereby dolphins scare fish out of protective cover with a loud tail slap, and **conching**, in which dolphins lift conches out of the water to manipulate the contents (that is, fish have been observed in the shells).

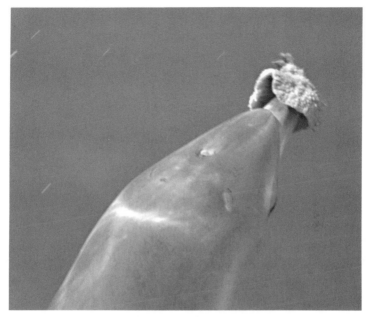

Figure 4.28. Sponging by bottlenose dolphin (courtesy of Janet Mann).

MATING AND SOCIAL SYSTEMS, REPRODUCTION, AND LIFE HISTORY
Promiscuity and Sperm Competition

All cetaceans mate and give birth in water. Most baleen whales are **promiscuous** breeders, mating with more than one partner during a single breeding season. When females have multiple mates, there is selection for males who can produce large amounts of sperm and therefore have larger testes. Males with larger testes are more likely to induce **sperm competition**, the physical competition among sperm from different males to fertilize the female egg.

Baleen whales do not group closely together or form large social groups. This probably reflects their need to feed more or less individually because small areas cannot sustain the large daily food requirements of a group. Another important factor is the seasonal availability of mysticete prey (for example, krill). The migration of some baleen whales is closely related to the availability of prey. The basic social unit is a mother-and-calf pair, and most mysticetes occur in small unstable groups of two to twelve, although only a few species have been studied in detail. A principal feature of the mating behavior of humpbacks is the performance by males of songs of extended duration, as described previously. This aggregation of displaying male humpbacks to attract females has been described as a lek.

Pods, Schools, and Family Units

Three general types of social groups have been described. The basic unit of a killer whale resident society is the **matriline,** a highly stable group of individuals based on maternal relationship. A typical matriline is composed of a female, her sons and daughters, and the offspring of her daughters. Several higher levels of social organization in resident killer whales exist, including the **pod,** a group of related matrilines that likely shared a recent common ancestor. The majority of pods are composed of one to three matrilines. All pods within a larger unit, a clan, have a similar vocal dialect. The top level of structure is the community, which is made up of pods that regularly associate with one another.

MALES ON THE RUN

During the breeding season, female humpbacks in estrus lead males into open water where males compete with one another for mating rights. As the chase ensues, the males try to impress the female by coming up out of the water or **breaching**, slapping their flippers against the water's surface, blowing bubbles, and singing. The male who wins access, the principal escort, attends the female. Other males, termed secondary escorts, attempt to displace the principal escort.

Many odontocetes, such as dolphins, live in **schools** characterized by long-term associations among individuals. School size is variable depending on species, location, and feeding habits. Some oceanic dolphins can occur in schools of hundreds or thousands of individuals. Schools also serve to protect individuals from predation.

Sperm whales occur in associations of varying sizes. They are sexually dimorphic, with males being considerably larger than females, a trait common in polygynous mating systems. Adult female and immature whales occur in **family units** consisting of about a dozen individuals and range throughout tropical and subtropical latitudes year-round. Like resident killer whales, units are structurally equivalent to pods and appear to be matrilineal. Young males leave their family unit at between 4 and 21 years of age and travel in loose associations or bachelor schools to higher latitudes.

Life History

SEXUAL MATURITY

The average age of sexual maturity in whales ranges in baleen whales from four years for humpbacks to approximately 10 years for sei, fin, and Bryde's whales and up to 25 years for bowheads. Among odontocetes, estimates of the age of sexual maturity range from three years for harbor porpoises to 14 years for killer whales. A peculiarity of whales useful to determine sexual maturity in females is that a past record of ovulation is retained for the animal's life. Each **corpus albicans**, a scar on the ovary, represents one ovulation (although not necessarily a pregnancy).

PREGNANCY AND BIRTH

Whales give birth to a single calf. Gestation ranges from 10 to 13 months for mysticetes and from 7 to 17 months for odontocetes. Given the large energy demands on breeding females, most cetaceans do not breed every year, although some smaller species, such as the minke and harbor porpoise, do breed annually. Some larger whales, such as blue and killer whales, have calving intervals every three years.

Life history data, including estimates of population sizes, can provide important conservation information. For example, although current population sizes for several species of large whales have increased, they are still a fraction of prewhaling population sizes. Knowing this could help determine an effective conservation strategy for these whale populations. Another example of the importance of life history data is the increase in births of bottlenose dolphins in the Mississippi Sound since Hurricane Katrina, which has been attributed to reduced fishing and boat traffic. Further monitoring is necessary to determine whether this new higher population level is sustainable.

PARENTAL CARE AND LACTATION STRATEGIES

Cetacean milk has a higher fat content (30–50 percent) than any other mammal except pinnipeds. Milk is forcefully ejected since neonates must hold their breath during underwater nursing. The duration of lactation varies from six to seven months in baleen whales, for whom food is seasonally abundant and relatively easily caught and eaten to a year

HOW DO WE KNOW HOW OLD THEY ARE?

The age of odontocetes can be determined by counting growth layers in teeth (see chapter 3). In mysticetes, the large waxy ear plugs, up to a meter (3 ft) long, when sectioned, contain layers that represent annual increments. However, reliability is questionable given the difficulty of interpreting layers, particularly after the onset of sexual maturity. Age estimates of bowhead whales based on changes in eye lens proteins suggest that they may be the longest-lived mammals known, living more than 200 years! Supporting evidence for this age estimate comes from points of ivory harpoons, which have not been used since the 1880s, found in the blubber of whales killed over the last four decades.

or longer in some larger odontocetes (for example, sperm whales), for whom food is more dispersed and difficult to locate and capture.

As noted previously, most baleen whales are characterized by long migrations between the winter low-latitude breeding grounds and high-latitude summer feeding grounds. Thus mysticetes exemplify the fasting strategy similar to phocids (see chapter 3), feeding little if at all during their winter migration. Toothed whales are characterized by an aquatic nursing strategy and most females feed throughout lactation.

GROWTH AND LIFE SPAN

The fetus in baleen whales grows faster than in any other mammal. During the first few months after birth, the young can increase its weight by five to eight times. Female mysticetes are generally slightly larger than males and some odontocetes, such as sperm whales, also exhibit sexual dimorphism in which males are considerably larger than females. The low reproductive rate of cetaceans is offset by the fact that they are long lived. Among whales, bowheads likely live the longest at more than 100 years. Bottlenose dolphins live for 50 years in the wild.

CHAPTER FIVE

Diversity, Evolution, and Adaptations of Sirenians and Other Marine Mammals

Sirenians or sea cows include manatees (Trichechidae), comprising three species, and the Dugongidae, with a single species, the dugong. Sirenians derive their name from mermaids of Greek mythology, known as sirens. Although clearly not the original sirens discussed in the classic poem *The Odyssey*, as the Greeks are unlikely to have encountered manatees or dugongs, early accounts of Columbus on a voyage in the Caribbean in 1493 described manatees as "sirens." Sirenians are unique among living marine mammals by having an herbivorous diet, feeding almost entirely on aquatic plants. Like whales, they are totally aquatic but their ancestors had four limbs and moved on land in the past. Sirenians were more diverse in the past and only two genera and four species survive today. The dugong (*Dugong dugon*) is found in the South Pacific and the Indian Ocean. Manatees (*Trichechus* sp.) occupy tropical and subtropical regions in the Atlantic Ocean and some of its major river drainage systems. Relationships among extinct and living sirenians are shown in fig. 5.1.

In addition to the three major lineages of marine mammals—pinnipeds, cetartiodactylans, and sirenians—there are several lineages of mammals with marine representatives, including sea and marine otters and polar bears (included within the Carnivora) and extinct aquatic sloths in Edentata.

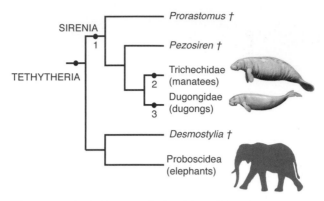

Figure 5.1. Evolutionary relationships of sirenians. Selected shared derived characters of the groups identified by numbers.

1. **Sirenia:** External nares retracted and enlarged, reaching beyond anterior margin of eye orbit; premaxilla contacts frontal bone; absence of sagittal crest; thick, dense skeletons.
2. **Trichechidae:** Reduction of neural spines on vertebrae; "conveyor belt" tooth replacement; paddle-shaped tail.
3. **Dugongidae:** Strongly downturned snout; incisor tusks (males); notched tail.

WALKING SEA COWS!

Sirenians have a fossil record that extends from the early Eocene (50 Ma) to the present. Proboscideans (which include elephants) are the closest living relatives of sirenians. Sirenians, proboscideans, and extinct desmostylians are recognized as a clade termed **Tethytheria**, so named because early members were thought to have inhabited the shores of the ancient Tethys Sea (see also chapter 2). Tethytheres, together with elephant shrews, tenrecs, golden moles, and hyraxes have been grouped into the larger **Afrotheria** clade, named for the African origin of its members.

The earliest known sirenians are prorastomids from early and middle

Figure 5.2. Reconstructed composite skeleton of *Pezosiren portelli*; unshaded areas partly conjectural (see Berta et al. 2006 for original data source).

Eocene rocks (50 Ma) of Jamaica. The dense, swollen ribs of prorastomids indicate a partially aquatic lifestyle, as does their occurrence in lagoonal deposits. *Prorastomus sirenoides* was approximately 1.5 m (5 ft) in length. Unlike modern sirenians but similar to early cetaceans, prorastomids possessed well-developed legs and were capable of locomotion both on land and in the water. Judging from their crown-shaped molars, they likely fed on soft plants. *Pezosiren portelli*, a pig-sized animal, possessed an elongated rostrum and a long trunk supported on four relatively short legs (fig. 5.2). *Pezosiren* occupied both land and water and swam, like otters, by spinal and hind limb undulations aided by a long tail—a convergence they shared with early whales.

Study of the skull of *Protosiren fraasi*, another early sirenian, using CT scans revealed that this animal possessed small olfactory (smell) and optic (vision) tracts but large maxillary nerves, consistent with the diminished importance of smell and sight in an aquatic environment. The enlarged, downturned snout of most sirenians confirms their enhanced tactile sensitivity. Oxygen isotope studies of their teeth show that protosirenids were exclusively seagrass feeders. In contrast to cetaceans, which first evolved in freshwater, passed through a variety of amphibious forms, and then rapidly evolved into fully aquatic animals, sirenians first exploited marine ecosystems and only later in their evolution invaded freshwater.

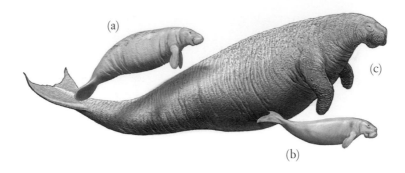

Figure 5.3. Sirenians: (a) West Indian manatee, (b) dugong, and
(c) Steller's sea cow (painted by Carl Buell).

CROWN SIRENIA

TRICHECHIDAE (MANATEES)   Living manatees include three species
that are mostly tropical in distribution: West Indian manatee (*Trichechus
manatus*), West African manatee (*Trichechus senegalensis*), and Amazon manatee (*Trichechus inunguis*). The generic name of the manatee, *Trichechus*,
comes from the Greek and refers to sparse body hair and abundant facial
hairs and bristles. Manatees differ from the dugong in having a rounded
fluke, continuous tooth replacement, and a less sharply deflected snout
(fig. 5.3).

The earliest manatee was *Potamosiren magdalenensis* from the Miocene (15 Ma) of South America. The evolution of manatees was tied
to geologic events, specifically the increased abundance of freshwater
grasses in Amazon river systems triggered by the uplift of the Andes
Mountains. Manatees evolved small, numerous teeth with complex
chewing ridges, enabling them to more efficiently process aquatic
grasses that were higher in silica content and therefore more abrasive.
The Andean uplift isolated the Amazon basin from the Pacific. Manatees trapped in this region evolved into the Amazonian manatee. The
West Indian manatees on the Atlantic coast of South America evolved
into the Florida and Antillean subspecies. The West African manatee

appears to have diverged from West Indian manatees when they dispersed to Africa 1.5 Ma.

DUGONGIDAE (DUGONGS)   Living dugongs are represented by a single species, *Dugong dugon,* and inhabit coastal waters from eastern Africa along the southern edge of Asia to the eastern coast of Australia. With the exception of Australian waters, most populations are small and fragmented. For the most part, they occur in nearshore waters. Dugongs can be distinguished from manatees in having a dorsally notched tail like cetaceans, incisor tusks (in males), and a sharply downturned snout (fig. 5.3).

Early dugongs were widely distributed and are known from North America, Eurasia, the Caribbean, and South America. Later diverging taxa (for example, *Dioplotherium* and *Rytiodus*) possessed well-developed, blade-like tusks presumably used in feeding. Smaller tusks in the living dugong likely represent an exaptation, given their use in male-to-male encounters.

The extinct hydrodamaline lineage, made up of large-bodied animals, includes the recently extinct Steller's sea cow (*Hydrodamalis gigas,* fig.5.3). The Steller's sea cow, named for its discoverer, the German naturalist George W. Steller, was a gigantic animal, measuring 7.6 m in length with an estimated weight of 4–19 tons. This sea cow was unusual in lacking teeth and finger bones and in possessing thick, bark-like skin. The large size of the Steller's sea cow was an adaptation to living in cold waters near islands in the Bering Sea, in contrast to the distribution of other sirenians in tropical or subtropical waters. It swam by dorsoventral undulations of its body and its horizontally expanded tail. Steller described the sea cow's blubber, 8–10 cm (3–4 in) thick, as tasting something like almond oil. Unfortunately, the Steller's sea cow quickly became a major food resource for Russian hunters. By 1768, only 27 years after its discovery by Europeans, the sea cow went extinct. Scientists modeling the extinction of the Steller's sea cow based on historical records and life history data suggest that the animals were greatly overexploited, hunted at seven times the sustainable limit. This further suggests that the initial sea cow population was relatively small, an estimated 2,900 animals.

EVOLUTIONARY TRENDS

The earliest sirenians were moderately sized four-legged animals of nearshore marine habitats. Beginning in the late Eocene with the evolution of dugongs, they had become completely aquatic with a streamlined body, well-developed foreflippers and a powerful, horizontally expanded tail. The progressive loss of hind limbs in sirenians is well documented in the fossil record. A heavy skeleton was present early in sirenian evolutionary history and suggests that shallow diving was an early adaptation. Sirenians have since declined from their peak diversity in the middle Miocene (12–14 Ma), the result of a combination of factors including climate change, loss of habitat, and human activity.

STRUCTURAL AND FUNCTIONAL INNOVATIONS AND ADAPTATIONS
Dense Bones and Buoyancy

The bones of sirenians are dense and lack marrow cavities, a condition referred to as **pachyosteosclerotic**. Dense bones act like weight belts used by human divers and help sea cows stay submerged. Other adaptations, such as the horizontally positioned diaphragm and lungs, facilitate buoyancy by changing the volume of air in the lung cavity. Contractions of the muscular diaphragm and abdomen may also compress gas in the large intestine and contribute to buoyancy control.

How Manatees and Dugongs Lost Their Legs

Like whales, living sirenians lack hind legs. A study of loss of pelvic fins in species of three-spined stickleback fish and the observation that sirenians and sticklebacks share asymmetry in pelvic bones (elements on the left side are larger than those on the right side) led researchers to suggest that similar genetic mechanisms may underlie pelvic reductions in both groups. Gene expression studies showed that a mutation in *PitX1* gene was responsible for pelvic reduction in several species of sticklebacks and hind limb loss in laboratory mice. Although more study is needed to link this mutation with hind limb loss in manatees and dugongs, it supports other evidence that particular genes have evolved repeatedly and independently.

Flippers and Flukes

The forelimbs are primarily used in steering and orientation, although they are employed in propulsion when the animal is in contact with the sea floor. The elbow joint is mobile, as are the wrist and finger bones. Although the West Indian manatee possesses nails at the tips of each digit, nails are absent in Amazonian manatees.

Manatees differ from dugongs in having a rounded, paddle-shaped

---

SWIMMING

Sirenians, like cetaceans, swim by up-and-down undulations of the tail. They are slow swimmers. Although clocked at speeds of up to 15 mph for short bursts, manatees generally cruise at speeds of 2–6 mph (fig. 5.4).

Manatees differ from virtually all other mammals in having fewer neck vertebrae; their shortened necks result in limited head mobility, which functions to decrease drag and aid in locomotion.

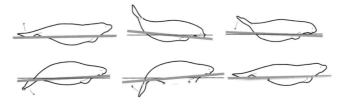

Figure 5.4. Lateral view of a West Indian manatee swimming; tracings of body, limb, and tail movements (modified from Hartman 1979).

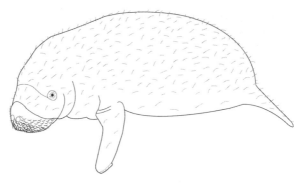

Figure 5.5. Hypothesized lateral line system of manatees (courtesy R. Reep).

tail, whereas dugongs possess a triangular tail like those of whales (see fig. 5.3). As in cetaceans, the tail of sirenians is composed mostly of dense connective tissue. Anatomical studies have revealed that the tail and flippers function as countercurrent heat exchangers to regulate heat loss (see chapter 3), similarly to those in cetaceans and pinnipeds.

SENSES

Body Hairs—A Lateral Line?

Manatees and dugongs possess numerous sensitive body hairs over their entire bodies (fig. 5.5). For manatees, this number averages around 3,000 body hairs. This type of body hair is normally present only on the muzzle (see chapter 3). These sensory body hairs consist of specialized cells associated with nerve cells that are stimulated with water movement. These body hairs appear analogous to the **lateral line** systems of fish and some amphibians, which are used to detect underwater objects. This capability would be useful for detecting topographic features, such as shorelines and sandbars, as well as approaching animals and needs further study.

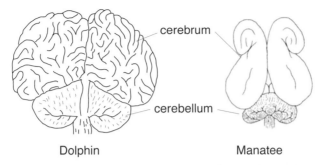

cerebrum

cerebellum

Dolphin   Manatee

Figure 5.6. Comparison of a dolphin and a sirenian brain (illustrated by P. Adam).

Temperature Regulation

Manatees have thick skin, as much as 2.5 cm (1 in) thick, but they do not possess a true blubber layer. Manatees, especially juveniles, are sensitive to temperature changes and migrate into warmwater when the temperature drops below 20 °C (68 °F).

Small Brains

The brain of sirenians is relatively small for their body size, with a smooth surface (fig. 5.6). This lack of folding on the cerebral surface is unusual for mammals. The low EQ of sirenians has been correlated with their low metabolic rate, simple social structure, and herbivorous diet.

Poor Sight and Hearing

The eyes of sirenians are small and spherical. Vision in air and water is generally quite good. The lack of well-developed ciliary muscles, which allow for close focusing, is consistent with their poor near vision.

Manatees and dugongs produce low-frequency sounds (3–5 kHz) described as chirp-squeaks. These sounds are produced in the larynx and most are thought to function as mother—calf vocalizations. It has been suggested, although not confirmed, that the main area of sound reception is not through the small ear openings located behind the eyes but by thick, oil-filled cheek bones that are in direct contact with the ear bones. Hearing tests on manatees indicate that slow-moving boats don't provide a warning sound of the boat's path. This, coupled with their poor near vision, makes manatees especially vulnerable to boat collisions.

Taste and Smell

In comparison with other marine mammals, sirenians possess numerous, well-developed taste buds and associated mucous glands that produce enzymes that break down sea grasses and algae. Olfactory bulbs are small, suggesting a limited sense of smell.

DIET AND FEEDING SPECIALIZATIONS
Mobile Lips for Grazing on Sea Grasses

Sirenians are the only marine mammals that use their prehensile snout in feeding. The degree of snout deflection is correlated with degree of specialization for bottom feeding. The snout is more strongly downturned in dugongs, making this animal an obligate bottom feeder on sea grasses. Manatees have only a slight deflection, have a more generalized diet, and are able to consume floating aquatic plants (fig. 5.7).

The upper lip, also termed the **rostral** or **oral** disc, is very flexible and covered with stiff hairs that bring vegetation to the mouth. During feeding, dugongs gouge tracks, called feeding trails, into bottom sediments. Dugongs specialize in feeding on the rhizomes (stem portion) of sea grasses, presumably because they contain more starch than the leaves and are higher in nitrogen, required by all animals for the manufacture of protein.

Thalassia

Halodule

Syringodium and *Thalassia*   *Halophila*

Figure 5.7. Examples of aquatic plants eaten by sirenians (courtesy Keys Marine Lab).

Teeth

Although male dugongs develop incisor tusks, they do not appear to be used in feeding but rather in fighting. Some fossil dugongines, however, possessed larger, more blade-like tusks that were likely used to dig up sea grasses. Most sirenians possess peg-like molars located in the back of the jaws, and chewing also occurs by grinding plates in front of the teeth on the roof of the mouth (fig. 5.8). The Steller's sea cow lacked teeth but possessed grinding plates for feeding on algae. Manatees possess a conveyor-belt dentition, where as worn teeth fall in the front they are replaced by new teeth erupting in the rear of the jaw, which is likely related to their ability to eat abrasive grasses that promote tooth wear. Unlike manatees, dugongs only replace their teeth once in their lifetime, a factor that may have contributed to their extinction in the Amazon basin and Caribbean.

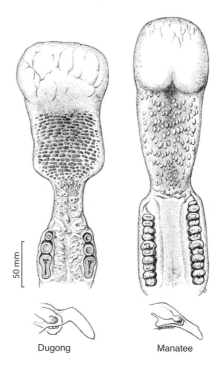

Figure 5.8. Comparison of grinding plates and snout deflection in a dugong and manatee (Marsh et al. 1999, with permission from Wiley and Sons, Inc.).

50 mm

Dugong

Manatee

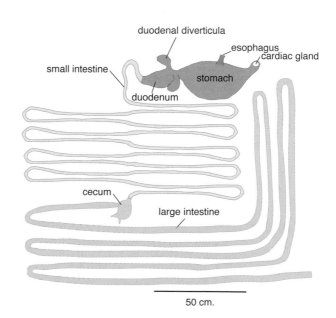

Figure 5.9. Diagram of a manatee digestive tract, drawn to scale (adapted from Reynolds and Rommel 1996).

duodenal diverticula

esophagus
cardiac gland

small intestine

stomach

duodenum

cecum

large intestine

50 cm.

Digestive System

Sirenians possess extremely long digestive tracts (their intestines measure more than 40 m or 130 ft in length, compared to humans at 9 m or 30 ft; fig. 5.9). They differ from ruminant digesters, such as cetaceans; rather, they are hindgut herbivores like horses, with a large intestine that functions as a fermentation chamber for the digestion of cellulose. Since hindgut digesters do not regurgitate, plant materials tend to be completely broken down by the time they pass through the intestines. Hindgut digesters are not as efficient as ruminants at extracting nutrients from food and, as a consequence, they feed constantly and are able to survive on poor-quality food, meaning that it is high in fiber, low in protein, and low in caloric value.

Water Balance

Of the three manatee species, only the Amazonian manatee lives exclusively in freshwater. The West Indian and West African manatee live in both freshwater and saltwater, which is also true for some pinnipeds, such as harbor seals, California sea lions, and some cetaceans, including the Amazon river dolphin and the Ganges river dolphin.

Any organism that lives in water must balance the difference between its internal fluid composition and the marine or freshwater medium that it occupies. Marine mammals manage their salt and water balance in several ways. This is known as **osmoregulation**. Some

Figure 5.10. Comparison of reniculate (lobed) versus nonreniculate kidney.

marine mammals (cetaceans, pinnipeds, and sea otters) have specialized or **reniculate** kidneys that increase the surface area for salt excretion (fig. 5.10). Other marine mammals, such as sirenians, produce more concentrated urine that enables them to reclaim more of the water that would otherwise be lost through evaporation across the skin or through respiration. Although some marine mammals are known to drink seawater on occasion, they obtain water from their food and by producing it internally from the metabolic breakdown of food.

MATING AND SOCIAL SYSTEMS, REPRODUCTION, AND LIFE HISTORY
Mating Herds

Manatees and dugongs are solitary for most of the year. Social interactions, apart from mother–calf interactions, are mainly focused on mating. Mating and calving can occur throughout the year, but mating herds form more often in the summer. Female manatees and dugongs are promiscuous, mating with multiple partners. Male dugongs differ from manatees in their more intense physical competition for females. In some locations, **mating herds** are observed composed of males (as many as 18) that surround a female in estrus, or heat, pushing and shoving for access to her. In other locations, such as Shark Bay, Australia, male dugongs appear to defend territories and are thought to exhibit a lek strategy.

Life History
SEXUAL MATURITY

Female manatees reach sexual maturity at between 2.5 and 6 years of age, with males reaching sexual maturity at between 2 and 11 years. Female and male dugongs mature much later at 9.5–17.5 years. The longer time for sexual maturity in males has been related to their need to reach a certain body size to successfully engage in male-to-male competition to gain priority access to females in a mating herd. However, this has important conservation implications, since males may die before they reach sexual maturity.

PREGNANCY AND BIRTH

Gestation period for manatees and dugongs is approximately one year, varying from 12 to 14 months. Manatees reproduce every two to three years; however, in dugongs this number is higher, ranging from three to nearly six years. This low reproductive rate leaves them vulnerable to decline.

PARENTAL CARE AND LACTATION STRATEGIES

For both manatees and dugongs, males have no role in parental care. Sirenians exhibit an aquatic nursing lactation strategy. Manatee calves stay with their mother for at least one year. Lactation is relatively long, lasting at least 1.5 years. Compared to pinnipeds, the milk of sirenians contains less fat (13 percent).

GROWTH AND LIFE SPAN

Average life span for manatees and dugongs is similar—approximately 30 years—although there are individuals with age estimates of 60 years or more. The age of male dugongs can be determined from growth lines in their tusks. Since the teeth of manatees are replaced continuously, annual growth layers in the ear bones provide a more reliable age estimate.

DESMOSTYLIANS

Named for their bundled columnar tooth cusps, desmostylians constitute the only extinct order of marine mammals. They are related to sirenians and elephants and grouped in the Tethytheria clade. Seven genera and 12 species are grouped into two families: Paleoparadoxiidae and Desmostylidae. They were hippo-sized, sexually dimorphic quadrupeds that originated in Asia along the ancient Tethys Sea and migrated to the Pacific coast approximately 30 Ma, where they were confined to coastal

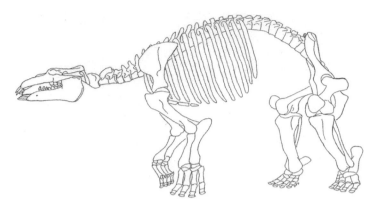

Figure 5.11. Restored skeleton of desmostylian *Paleoparadoxia tabatai* (see Berta et al. 2006 for original source).

marine rocks ranging as far south as Baja California, Mexico. Most had a long, narrow, slightly deflected rostrum. Reconstructions of their skeletons and inferred locomotion are debated, although it appears that they had upright posture when moving on land and that they were propelled by the forelimbs in the water (fig. 5.11). Also controversial is the diet of desmostylians. At one time, researchers hypothesized a diet of mollusks for demostylians based on their heavy columnar teeth. Their high crowned teeth, however, are more similar to those of grazing ungulates, which, together with stable isotopic analysis of the teeth (see chapter 1), suggests a diet of aquatic plants. *Desmostylus* was an aquatic herbivore that spent a considerable portion of its life foraging in estuarine and freshwater ecosystems. Given their limited geographic distribution, they were apparently adapted to cooler climates than the tropical sirenians but were likely outcompeted by the hydrodamaline sirenians.

AQUATIC SLOTHS

As many as five species of extinct marine sloth (*Thalassocnus*) have been described based on abundant, associated complete and partial skeletons. Each species is known from a different level of rock unit and their

Figure 5.12. Life restoration of aquatic sloth *Thalassocnus* (painted by Carl Buell).

progressive and differing aquatic adaptations suggest that they may actually represent a single evolutionary lineage. This marine sloth lineage was very restricted geographically, apparently endemic to Peru, and lived during the late Miocene–early Pliocene. The tail was used in swimming and the crushing, worn molars and downturned rostrum with an expanded tip suggest the presence of a well-developed lip for grazing on sea grasses (fig. 5.12). Aquatic sloths lack pachyostotic skeletons, found in other aquatic bottom feeders, and instead may have used enlarged claw-like forelimbs to anchor themselves on the bottom, perhaps similar to the Steller's sea cow. The similarity of these marine sloths to extinct desmostylians in the North Pacific raises the possibility that they may be the ecologic homologues of desmostylians in the South Pacific. Their extinction has been related to a decrease in sea temperature, which may have altered their food sources and/or availability.

MARINE OTTERS

Marine otters include the sea otter and a marine species of South American otter. These members of the carnivore family Mustelidae also include river otters, weasels, and badgers. Although sea otters are the smallest marine mammals (1.4–1.5 m or nearly 5 ft in length, weighing 45 kg or 31–99 lb), they are the largest mustelid.

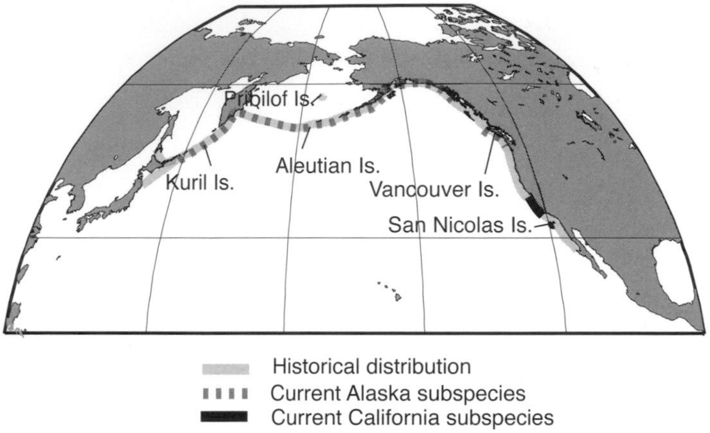

Figure 5.13. Sea otter distribution map (from Estes and Bodkin, 2002).

Three subspecies of the sea otter (*Enhydra lutris*) are recognized, based on differences in morphology and geographic distribution: the common sea otter (*Enhydra lutris lutris*) inhabits islands off Japan along the Pacific rim; the northern sea otter (*Enhydra lutris kenyoni*) ranges from Alaska to Oregon; and the southern sea otter (*Enhydra lutris nereis*) had a historic range from northern California to central Baja California, Mexico, and is present today along the central California coast (fig. 5.13).

The closest living relatives of *Enhydra* are other lutrine otters (that is, *Lutra*, *Amblonyx*, and *Aonyx*). The giant fossil otter (*Enhydritherium*), known from the Miocene of Europe and the late Miocene/Pliocene of North America and a close relative of the sea otter, was adapted to freshwater as well as coastal marine environments. *Enhydritherium* differed from *Enhydra* in being a forelimb-dominated swimmer and in having skeletal specializations (that is, more equally proportioned fore-limbs and hind limbs), suggesting that its locomotion on land was more proficient.

Marine otters (*Lontra felina*) are rare and poorly known marine mammals. They are more distantly related to *Enhydra* and other lutrine otters. They are relatively small, approximately 1 m (3 ft) in length.

They are the most exclusively marine species of the otters of South America and rarely venture into freshwater or estuarine habitats. They mainly inhabit rocky shorelines of southern Peru, the entire coast of Chile, and the extreme southern reaches of Argentina, where there is abundant seaweed and kelp, and infrequently visit estuaries and freshwater rivers.

Evolutionary Trends

Although at the present time marine otters tend to live in higher latitudes, this was not true in the past, given the much broader latitudinal range of fossil otters. Fossil evidence indicates that otters have secondarily entered the sea from a freshwater origin. It has been suggested that the higher food availability and productivity of coastal marine habitats compared to freshwater systems may have driven their evolution from freshwater into the sea. Supporting evidence for the evolutionary success of marine otters includes their high population density in marine environments compared to the low density of freshwater otters.

Structural and Functional Innovations and Adaptations

Unlike most marine mammals, the primary form of insulation for sea otters is an exceptionally thick coat of fur, the densest of any known animal. A sea otter's fur is actually made of two layers: long outer guard hairs that provide protection for the fine dense **underfur**. One square inch of sea otter underfur contains between 170,000 to over 1,000,000 hairs, more than the hairs on a human head! Their thick fur was prized by hunters in the past and resulted in the near extinction of the species. When immersed in water, the underhairs overlap and help maintain trapped air and reduce heat loss. To maintain the insulating quality of their fur, sea otters spend considerable time and energy grooming.

Although sea otters are able to walk on land, they spend most of their time in water resting in kelp beds. They are primarily hind limb—dominated swimmers feeding on various invertebrates, notably sea urchins. Sea otters generally swim belly-up, paddling with the webbed hind flippers. Neither the tail nor the forefeet play a role in propulsion. Sea otters swim relatively slowly at about two or three miles per hour (0.6–1.4 m/sec) Sea otters haul out on land relatively infrequently, and their movement on land is awkward and clumsy.

Sea otters are an important component of nearshore marine communities, especially kelp forests (see also chapter 6). Sea otters feed on bottom-dwelling invertebrates, especially sea urchins, mollusks, and crustaceans. Otters need to consume at least 25 times their body weight in order to stay warm. Their method of feeding is unique. Captured prey is manipulated with the forepaws or is temporarily held in loose skin pouches in the armpits. Sea otters and some bottlenose dolphins (see chapter 4) show the use of tools. For larger, heavier shelled prey, sea otters will sometimes carry a rock to the surface, place it on the belly, and use it to crack open the hard shell while they float on their backs.

Mating Systems and Social Structure

Sea otters have a polygynous mating system. Marine otters are reportedly monogamous unless additional potential partners are available. Mating occurs year-round. Adult male sea otters maintain territories related to the distribution of kelp, with adult females moving freely among these territories. After copulation, males separate from females.

Researchers have tracked sea otters and discovered that, seasonally, males travel 50–60 mi or more to male groups at either end of the range. In comparison, females are sedentary and generally remain in a relatively smaller home range territory. Sea otters tend to rest in groups called rafts. These rafts can consist of just a few otters to 40–50 otters in California to groups as large as 2,000 otters in Alaska.

Similar to pinnipeds and the polar bear, sea otters undergo delayed implantation, which ensures that young are born during a favorable time of food availability, generally in the early summer. Sea otters bear a single pup. Marine otters bear a litter of two to five pups. Birth occurs in water. Sea otters in different geographic areas have variable birth interval rates, which depend on season and food availability. Generally, California otters give birth every year, whereas otters in Alaska give birth every two years. Sea otters employ an aquatic nursing strategy and sea otter milk is very rich in protein and fat (as much as 25 percent fat). Like other polygynous species, male sea otters provide no parental care. Females have been observed caring for orphaned pups.

Male sea otters live for between 10 and 15 years, while the life span of females is longer, approximately 15–20 years.

POLAR BEARS

The polar bear (*Ursus maritimus*) the world's largest bear, occupies the Arctic Ocean and its surrounding seas. An adult male weighs around 400–680 kg (882–1,411 lb), while an adult female is about half that size. The forepaws are especially large and paddle-like, presumably an adaptation for swimming and for walking on ice and snow. Polar bear fur is white, yellow, or gray or brown, depending on the season and lighting condition. The green coloration sometimes seen in captive polar bears is due to growth of algae in the hollow hair shaft.

Previous molecular data suggested that polar bears diverged from brown bears between 1 and 1.5 Ma. DNA analyses of remains of the oldest known polar bear, found north of the Arctic Circle, provide an evolutionary snapshot revealing when and where polar bears evolved. This new data confirms that polar bears originated in the Arctic but provides evidence that they split more recently from brown bears, probably less than 500,000 years ago. Stable isotope analyses of the carbon and nitrogen in the bear's tooth indicate that polar bears clearly had a different feeding ecology from that of brown bears and that, early in their evolutionary history, polar bears had adapted rapidly to feeding on seafood, granting them their unique position as top marine predators in the Arctic. As discussed earlier (chapter 1), genetic evidence contradicts fossil data and suggests that polar bears originated near Britain and Ireland.

Structural and Functional Innovations and Adaptations

Polar bears are efficient swimmers. The forepaws provide most of the propulsion during swimming, with the hind limbs trailing behind. Like humans, polar bears walk on the soles of the feet, a **plantigrade** foot posture. On land, the robust limbs and plantigrade feet help distribute the body weight. The foot pads possess tiny bumps that increase friction between the feet and ice.

The principal prey of polar bears is the ringed seal (*Pusa hispida*) and to a lesser extent, the bearded seal (*Erignathus barbatus*). A key adaptation of polar bears is the ability to store large amounts of fat when prey is available and then to fast for long periods of time. Pregnant females are without food the entire period in the den (up to eight months). Males and juveniles of both sexes remain on the ice and hunt throughout the winter.

Mating Systems and Social Structure

Polar bears are sexually dimorphic and they have a polygynous mating system. They are generally solitary, although females may remain and travel with their cubs for several years. Mating generally occurs from late March through early May. As is typical of polygynous mammals, female polar bears reach sexual maturity earlier than do males.

Life history Females reach sexual maturity when they are three to four years old, with males slightly later at five to six years. Gestation in polar bears lasts 5.4 to 8 months. Differing from all other marine mammals, polar bears typically give birth to twins, although three cubs are

Figure 5.14. Polar bear cubs in snow shelter (courtesy G. Thiemann).

not uncommon. One of the effects of global warming and the reduction of sea ice is that litter size declines, a serious threat to the population viability of polar bears (see chapter 6). Pregnant female polar bears enter maternity dens in snowdrifts (fig. 5.14). Polar bear cubs are born in December or January in a very underdeveloped **altricial** state and remain confined to the den, consuming fat-rich milk for two to four months following birth. The average life span of polar bears is 20–25 years.

CHAPTER SIX

Ecology and Conservation

Broad aspects of the biology of marine mammals are correlated with their ecological role—that is, how they make a living—and respond to the distribution of resources (for example, food and territory) and competitors in a community. This ecological role, explored in this chapter, includes food and the feeding relationship of marine mammals and the effects of change on the dynamics of marine ecosystems. Also considered is the influence of humans on marine mammal communities, ranging from oil spills and contamination of the ocean with heavy metals to diseases and global climate change and loss of habitat. Finally, conservation efforts such as captive breeding, provision of critical habitat, and relocation and recovery programs are evaluated.

WHAT MARINE MAMMALS EAT
AND WHAT EATS THEM

Food chains or **webs** form the framework of most ecological communities, defining the energy flow through a chain of different organisms from extremely abundant primary producers (for example, phytoplankton) to relatively few predators (for example, animals). Most primary producers are photosynthesizers—that is, they use sunlight as their energy source. Nearly all food chains are based on photosynthesizers, although a few marine food chains (for example, whale fall communities discussed later) depend on bacterial chemosynthesizers, which use a chemical source of energy for production. Most animals are part of more than one food chain and eat more than one kind of food. These interconnected food chains form a food web. Sirenians are unique among marine mammals in feeding at the base of the food chain on sea grasses and algae. All other marine mammals are carnivores, consuming herbivores and other carnivores at higher trophic levels (fig. 6.1 and chapter 1).

The food web dynamics of ecosystems are regulated by "top-down" and "bottom-up" processes. Top-down refers to a top predator feeding on lower levels of the food web, whereas bottom-up refers to interactions in the opposite direction that are resource or prey driven. Sea otters and killer whales are part of marine food chains and act as dominant predators or **keystone species** that control species diversity in lower levels.

Sea Otters as Keystone Species

Sea otters in the North Pacific Ocean are considered keystone species, which means that what they eat has a large impact on the environment. In kelp bed communities they play a pivotal role by keeping populations of sea urchins, which in turn feed on kelp, low in number (fig. 6.2). In this way, sea otters promote the growth of vegetation—in this case, kelp—in much the same way that wolves benefit trees and shrubs by killing deer. The Australasia biotic assemblage in the Southern Hemisphere, by contrast, lacks a predator comparable to the sea otter. In such a food web, kelp is left unprotected and has evolved chemical defenses to protect against herbivory by sea urchins.

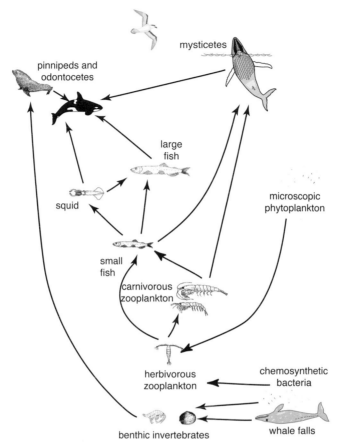

Figure 6.1. Food chain in a typical marine ecosystem (courtesy J. Sumich Whale Cove Education).

Killer Whales and Changing Ecosystem Dynamics

Killer whale predation is thought to be one of the major factors affecting populations of multiple marine mammal species. It is unclear to what degree killer whales preyed upon certain marine mammals during the onset of their declines, but recent evidence suggests that they may be a factor currently preventing species from recovering from their depleted status. To further complicate the issue, killer whales are declining in

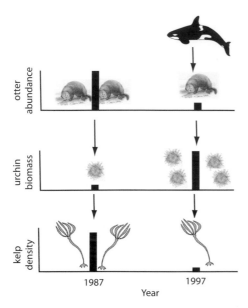

Figure 6.2. Comparison of ecosystem in coastal waters of Aleutian Islands in 1987 and 1997 where sea otters declined, sea urchins increased, and kelp decreased. Vertical bars are relative indices of variable listed. Note that 1997 sequential collapse is hypothesized to have been caused by killer whales "fishing down" the food web (adapted from Estes et al. 1998).

numbers where Steller sea lion and sea otter populations are in crisis, and abundant where Steller sea lions are showing signs of recovery.

An extension of the predation hypothesis is that, after peak commercial whaling, killer whales did not have the large baleen and sperm whales to prey upon as populations were decimated. This prompted a shift in predation to Steller sea lions in the early 1970s. This shift can be viewed as a **top-down** effect, which in ecology refers to a top predator—in this case, killer whales—controlling the ecosystem. The population decline of Steller sea lions and the slow recovery rate of cetacean populations caused killer whales to switch again, sequentially, to the other marine mammal populations that have declined recently, such as harbor seals, northern fur seals, and finally, sea otters. Evidence to support this hypothesis is the timing of the collapse, known diets, and observed foraging behavior of killer whales and their prey. The alternative **bottom-up** forcing proposes instead that overfishing and productivity are culprits, which in the case of Steller sea lions and other prey suggests that they have collapsed because their prey population (bottom-up) declined (see later discussion and fig. 6.3).

Decline of Steller Sea Lions and the "Junk Food" Hypothesis

The recent massive decline in the western stock of Steller sea lions (the population occupying the region from Alaska to the Aleutian Islands) has been attributed to a change in prey quality and quantity. Changes in the ecosystem of the North Pacific, likely due to both overfishing and global warming, have altered the availability of nutrient-rich salmon and herring. Instead, the sea lions are eating less nutritious cod and pollock, so-called junk food. Human junk food is fatty, but this junk food is the opposite—food without enough fat and energy to sustain them.

The junk food hypothesis was tested at the University of British Columbia in Vancouver, Canada. Scientists fed captive sea lions either pollock or herring. They found that adults could survive on pollock, but yearlings could not eat enough to sustain themselves. Other scientists are not convinced and argue that the sea lion decline was the result of other causes, such as their incidental catch by fisheries.

Whales and Southern Ocean Food Webs

Another example of the effects of top-down predator removal comes from the Southern Ocean ecosystem surrounding Antarctica, where removal of large whales is argued to have led to a krill surplus. According to this hypothesis, a krill surplus resulted in changes in predator populations, such as penguins and minke whales, that were released from trophic competition with large whales. Debate continues with both sides marshaling evidence to support either top-down or bottom up explanations.

Whale Fall Communities

Like the animal communities around deep-sea hydrothermal vents, **whale fall communities** evolve around food sources, defined as a dead whale sinking to the sea floor, where sulfides spewing from the vents are

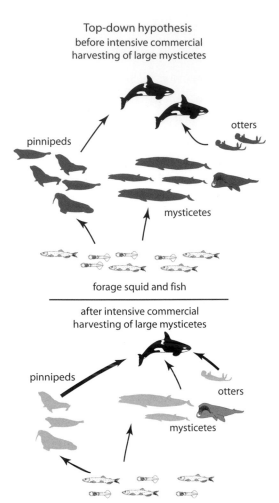

Figure 6.3. Marine mammal community illustrating top-down and bottom-up processes. Thickness of arrows indicates relative intensity of foraging by killer whales on two alternative prey types.

consumed by bacteria, which in turn provide nutrition for animals. This sulphophilic (sulfur-loving) whale fall community differs from typical food webs by being based on chemosynthetic bacteria instead of photosynthetic organisms. Roughly 10–20 percent of sulphophilic species found at whale falls are also found at hydrothermal vent communities, but the majority of the species found in each type of environment are unique.

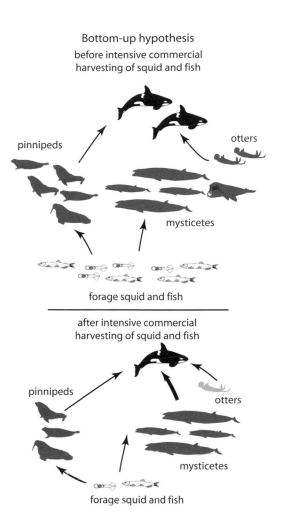

Bottom-up hypothesis
before intensive commercial
harvesting of squid and fish

pinnipeds

otters

mysticetes

forage squid and fish

after intensive commercial
harvesting of squid and fish

pinnipeds

otters

mysticetes

forage squid and fish

Whales as Carbon Sinks and Nitrogen Recylers

An intriguing study suggests that sperm and baleen whales may affect greenhouse gases in the Southern Ocean by taking up large amounts of carbon during feeding. Whales take up carbon just by bringing up to the surface squid (in the case of sperm whales); krill (for baleen whales); and other nutrients normally found in deep water, such as iron. The extra iron that whales bring up from their deep feeding encourages plankton growth and that growth traps carbon. Although

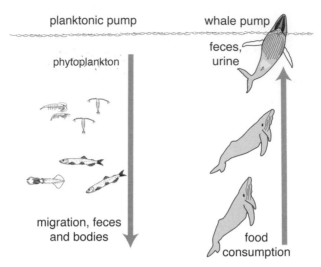

Figure 6.4. A conceptual model of a whale pump (adapted from Roman and McCarty 2010). In a typical planktonic pump, zooplankton and fish feed and export nutrients via sinking fecal pellets. Excretion for whales is likely shallower in the water column, where they feed, since they need to surface to breathe.

whales will never offset the tremendous amount of carbon that we humans produce by burning coal and fossil fuels and driving gasoline-powered cars, whales act as carbon sinks, a reservoir that accumulates and stores carbon, thus removing it from the atmosphere. This was especially true in the past.

It has also been suggested that whales may play an analogous role to upwelling by enhancing the nitrogen cycle that is releasing fecal material, which tends to disperse rather than sink. This release of nitrogenous material in the water creates an upward "whale pump," enhancing nutrient availability where whales gather to feed (fig. 6.4). A study of nitrogen flux in the Gulf of Maine suggests that the whale pump played a larger role before commercial hunting, when nitrogen recycling was likely more than three times atmospheric input.

INTERACTIONS BETWEEN HUMAN AND MARINE MAMMALS: LESSONS LEARNED

Large marine vertebrates, including whales, sea cows, and monk seals, are now functionally or entirely extinct in most coastal ecosystems. A compelling case has been made for overfishing as the critical driver of human disturbance in ecosystems. For example, historical records document the extensive hunting of manatees by early colonists in the Americas and dugongs by aboriginal people in Australia. This over-fishing increased the vulnerability of sea grass beds, which provide food and habitat for manatees and dugongs among other vertebrates, to recent events such as increases in sedimentation, turbidity, and disease. The disruption of sea grass ecosystems ultimately led to increasingly fragmented populations of once large herds of dugongs and manatees. This and other examples of changing ecosystem dynamics (for example, killer whales and sea otters discussed earlier) point up the importance of collecting historical data that document human exploitation of coastal resources for food and materials. This historical data serves as an important management tool for ecosystem restoration since it reveals that recent events (for example, climate change, pollution) are often symptoms of problems with deep historical causes.

Diseases and Pathogens: Prevention?

Disease is a frequent cause of death in marine mammal populations. Among more common infectious agents are bacteria, viruses, and biotoxins. The most serious infections are caused by **moribilliviruses**, which infect all marine mammal lineages and cause distemper in dogs and measles in humans. One of the largest-scale die-offs resulting from a moribillivirus was the death of 17,000–20,000 harbor seals of the North Sea and adjacent regions in the late 1980s.

Biotoxins such as **domoic acid**, a marine algae that causes shellfish poisoning, have also caused massive deaths in various species of pinnipeds, cetaceans, and manatees. Human deaths and serious illness have also been attributed to domoic acid. Although the term *red tide* is often associated with harmful algal blooms, scientists prefer the term harmful algal blooms since red-colored tides can be produced by nontoxic organisms.

Exposure to pathogens in some populations has been related to reduced genetic variability, increasing their vulnerability to other infectious diseases. Unfortunately, the usual treatment in humans to improve immunity to disease, a vaccine, is not an option for wild populations, although they have been effective in preventing the spread of disease in some captive situations.

Environmental Contaminants

Although the most hazardous contaminants, including hydrocarbons (for example, benzene and methane); polychlorinated biphenols (PCBs previously used in the manufacture of a variety of goods); pesticides (for example, DDT); and heavy metals (for example, lead and mercury) are banned in many countries, they continue to pose problems because they persist in the environment today. Environmental pollutants are causing various reproductive abnormalities in marine mammals. For example, in the late 1900s scientists discovered that polar bears near Svalbard and the Barents Sea had both male and female reproductive organs and also had high levels of PCBs in their blood and tissues. Marine mammals are particularly susceptible to toxic compounds, with their significant blubber layers acting as storage sites for these contaminants. The tendency for these toxic compounds to accumulate and increase in concentration through food webs is known as **biomagnification.** Fat-soluble substances are especially harmful since they accumulate in the tissues and are often transferred from mother to pup during nursing.

Catastrophic oil spills affect marine mammals when their oil-soaked fur loses insulation and buoyancy. Longer-term effects result from the ingestion of oil while grooming or feeding. Sea otters are especially vulnerable to oil spills. The second largest environmental disaster in recent U.S. history, the *Exxon Valdez*, spilled 11 million gallons of oil into Prince William Sound in 1989, killing an estimated 5,000 sea otters and numerous other marine mammals. Rescue efforts were undertaken and a sea otter recovery plan was developed that zoos and aquariums use today to rehabilitate stranded and oiled otters and other animals in minor oil spills. The damage caused by oil spills affects not only marine mammals directly but their food supply as the result of biomagnification.

Fisheries: Entanglements and Conflicts

The incidental capture or **by-catch** of marine mammals as the result of fishing activities is a serious global problem. Marine mammals are subject to entanglement and drowning in nets, lines, and traps of commercial fishing gear. The number of entanglements has been reduced by the use of acoustic alarms or **pingers** that warn animals about the presence of nets. However, it has been discovered that the increased level of noise associated with these devices may negatively impact some marine mammal populations and some species habituate to the sounds, using them as a "dinner bell" to locate food. There are a number of marine mammals that have been identified in feeding-related interactions with fisheries. Among the best-known examples is the southern sea otter (*Enhydra lutris nereis*). Sea otters prey on a variety of shellfish valued by commercial fisheries, including abalone and sea urchins. Recent southward movement of southern sea otters along the California coastline has enlarged the overlap between otter feeding areas and abalone fishing areas. Another detrimental impact of by-catch is to cetacean social groups, where individuals have been removed from family groups and mother–offspring pairs.

DOLPHIN AND TUNA FISHERY: SUCCESSFUL RECOVERY

In the eastern tropical Pacific (ETP), large yellow fin tuna swim with various species of dolphins, including pantropical spotted, spinner, and common dolphins. This ecologic association has resulted in the deaths of many dolphins trapped in nets of the tuna purse seine fishery. Purse seine nets are long walls of netting that are set and used to encircle and trap fish, since the bottom of the netting can be pulled closed like a drawstring purse. Since dolphins often associate with schooling tuna, they are herded and trapped in the nets. The number of dolphins killed since the fishery began in the late 1950s is estimated to be over 6 million animals, the highest known for any fishery. The by-catch of dolphins in the ETP tuna fishery has now been successfully reduced by more than 99 percent, but even at the present level of about 1,000 dolphins/year, it remains among the largest documented cetacean by-catch in the world (fig. 6.5).

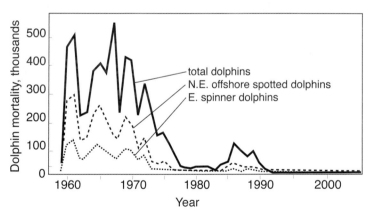

Figure 6.5. Annual dolphin mortality in the tropical Pacific Ocean tuna purse seine fishery, excluding numbers of calves separated from their mothers and presumed lost (redrawn from National Research Council 1992).

Mass Strandings: Still Mysteries?

Mass strandings are events that occur in many parts of the world in which groups of distressed cetaceans come ashore, strand themselves, and usually die on beaches. Pilot, beaked, and sperm whales and dolphins strand more often than other species. There are many possible explanations for why mass strandings occur. Several strandings in the past decade in the Canaries, Bahamas, and Hawaii have been related to military sonar, which may have caused temporary deafness resulting in beaked whales becoming disoriented and confused. Sonar may also be affecting the whales' dive patterns, causing them to stray into shallow water. Many beached whales in strandings associated with sonar also show evidence of physical trauma, including bleeding in their brains, ears. and internal tissues. For example, major organs of some examined beaked whales stranded in the Bahamas in 2000 showed tissue damage such as bubble formation within blood vessels characteristic of decompression sickness or "the bends," a condition that afflicts SCUBA divers who resurface too quickly after a deep dive. This suggests that these whales may be susceptible to gas bubble formation in the presence of intense sound pressures. In other cases, strandings are likely caused by weather conditions, diseases, and changes in underwater topography.

Global Climate Change and Its Impact: The Krill May Not Be Gone . . . But It Is Disappearing

Global warming is both a reality and, in large part, human caused. The increase of carbon dioxide and other greenhouse gases (so called because they trap heat) into the atmosphere is thought to be the main cause of global climate warming. Prediction models of the effects of global warming on patterns of marine mammal biodiversity have revealed local losses of native species (for example, polar bears) as well as some projected increases due to the invasions of temperate and subpolar species (for example, gray whales, Steller sea lions). Using global climate models, scientists can predict increases in temperature and receding ice. The impact of global warming on marine mammals has already been detected.

Increases in temperature have resulted in the disappearance of krill in polar regions, with some estimates indicating as much as an 80 percent loss of krill in the Antarctic in the last 40 years. The loss of sea ice removes a primary source of food for krill: algae that grows on ice. Given the significance of krill to the diets of many marine species, including crabeater seals and most baleen whales, its disappearance is alarming and cause for concern. The loss of sea ice also poses a threat to Arctic marine mammals like the walrus, polar bear, and hooded seal, which rely on sea ice as a platform for resting and reproduction. Walrus calves found swimming alone away from shore suggests mothers are forced to abandon pups with retreating ice. Polar bears may drown as a result of the large travel distances between receding ice sheets. Changes in ice thickness and extent also affect coastal habitat and the species that rely on it. The decline in sea ice has reduced the body size of polar bears and reduced their reproductive output and juvenile survival.

The warming of the ocean has changed species ranges, including migration routes and timing of migrations. For example, scientists have observed in recent years that gray whales delay their southward migration to the breeding lagoons off Baja California. As warmer waters melt the ocean's ice, other animals move into the whales' habitat and start feeding on the crustaceans. Crowded out by the new competition, the gray whales have to travel further north and feed longer to get their fill. These changes have altered the timing of the whales' yearly migration. A recent study by Pyenson and Lindberg, who analyzed the response of gray whales to climate change during the Pleistocene, suggests that the gray whale population in the eastern Pacific was larger than it is today. They attributed this, in part, to their generalist diet, which included herring and krill in addition to the benthic prey they consume today. This suggests that gray whales may have more adaptable diets than at present and therefore an increased ability to cope with climate change.

Changes in temperature have also changed the location of areas with high primary productivity. These areas are important to marine mammals because primary producers are the food source of marine mammal prey or are the marine mammal prey themselves. For example, while the numbers of blue whales feeding off the California coast have decreased in last 10 years, sightings of them have increased in the northern waters off Canada and Alaska. Research suggests that blue whales may be migrating north in search of krill, which has been depleted off California waters due to climate warming.

Noise

Marine mammals rely on underwater sound to navigate and hunt for food.

The potential effects of **anthropogenic** or human-generated noise on marine mammals continues to be a subject of controversy. Human-generated noise comes from many sources, including commercial ship traffic, seismic and oil exploration, acoustic research, and sonar. Observations of marine mammals indicate that anthropogenic sound can cause changes in their behavior and temporary or permanent injury and, in a few instances, can initiate events leading to death (see also section on mass strandings).

A positive step toward understanding the effects of undersea sound is occurring at Stellwagen National Marine Sanctuary at the mouth of Massachusetts Bay, where monitoring studies have documented how human activities are increasing undersea noise. Acoustic sensors have led researchers to assemble a "noise budget" for the preserve, which includes the calls of various marine mammals and fish as well as the sounds of nearby ships (fig. 6.6). The implications of this study are that larger ships produce sounds in the same frequency used by some marine mammals and fish, suggesting that their communication with one another may be seriously compromised by noise from shipping traffic.

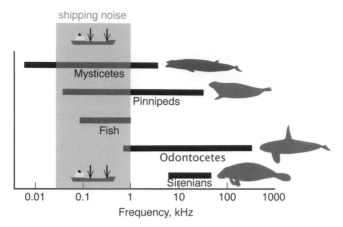

Figure 6.6. Sounds produced by marine mammals and ships in Stellwagen Marine Sanctuary, Massachusetts (modified from Science 2010).

Ecotourism: Whale Watching, Swimming, and Feeding

From tracking polar bears in the Arctic tundra to whale watching in California to swimming with dolphins in the Bahamas, ecotourist activities bring people in close contact with marine mammals. Unfortunately, such activities can be harmful to the animals since regulations governing such activities are sometimes weak or not strictly enforced. As a few studies have demonstrated, the presence of boat traffic and noise alters the behavior of marine mammals and they respond by increasing their swimming speed, changing their dive behavior, or leaving areas. Fortunately, some countries have established codes of conduct for observing marine mammals to be followed by tour operators as well as private boaters.

Other social interactions between humans and marine mammals include wild dolphin swimming and feeding programs. Although illegal in the United States by an amendment to the Marine Mammal Protection Act, such activities persist in other parts of the world. Again, studies have shown that often dolphins change their behavior when approached and, in some cases, aggressive behavior has resulted, placing the swimmer at risk. Feeding or provisioning wild marine mammals may also have serious consequences. For example, dolphins that become dependent on provisioning have been known to abandon their young, leaving calves to starve and fight off predators.

EXTINCTION: THE RULE, NOT THE EXCEPTION

Extinction is a normal process. Species have a natural duration from a few thousand to a few million years, so they live for a period of time and then disappear or become extinct. The most spectacular extinctions are **mass extinctions**, when a large number of different species die out relatively rapidly. At least five mass extinctions have been identified in the geologic past but only one has affected marine mammals, given their relatively recent evolutionary history. The current biodiversity crisis has been termed the sixth mass extinction, the only one that is human caused. Humans have accelerated the rate of extinction of many marine species, including marine mammals, by overfishing and hunting, habitat degradation, pollution, and global warming.

Can Biologists Predict Future Extinction Rates?

The rate of extinction of plant and animal species in the last 500 years is thought to be 1,000–10,000 times higher than it would be without the effect of humans. Assessing the probability of extinction of marine mammals is difficult as it necessitates determining the size of wild populations and changes in the populations during a specified time interval.

Conservation Status

Although the **Marine Mammal Protection Act of 1972** established a moratorium on the taking (and importing) of marine mammals, its jurisdiction is limited to U.S. waters. Another legislative act protecting marine mammals in the United States is the **Endangered Species Act (ESA)** of 1973. The International Union for the Conservation of Nature (IUCN), also known as the **World Conservation Union**, has summarized the best information available about the global conservation status of animals. The IUCN maintains a Red List of Threatened and Endangered Species. They place species in categories of risk by using criteria that focus on the absolute size of wild populations and changes in the populations in the last 10 years. According to the 2010 edition of the IUCN Red List, approximately one-third of marine mammal species are at great risk and are classified into the following categories: Extinct, Critically Endangered, Endangered, Vulnerable, Near Threatened, and Data Deficient (table 6.1).

EXTINCT

A species is extinct when no living individuals exist. Two modern species of pinnipeds are extinct: the Caribbean monk seal (*Monachus tropicalis*) and the Japanese sea lion (*Zalophus japonicus*). Although it is listed as critically endangered, recent reports have confirmed extinction of a single cetacean species: the baiji or Yangtze River dolphin (*Lipotes vexilifer*). Among crown sirenians, only the Steller's sea cow (*Hydrodamalis gigas*) is extinct, having the dubious distinction of becoming extinct only 27 years after it was discovered by Europeans.

CRITICALLY ENDANGERED

A species is critically endangered when it is facing an extremely high risk of extinction in the wild in the immediate future. Two pinnipeds, the Hawaiian monk seal (*Monachus schauinslandi*) and the Mediterranean monk seal (*Monachus monachus*), are listed in this category, along with a cetacean, the vaquita (*Phocoena sinus*). According to the latest reports, the vaquita is on the brink of extinction, with a population estimated at just 250 individuals, which is less than one-half of the number estimated about a decade ago.

TABLE 6.1

Numbers of species of marine mammals that are listed as extinct, endangered, vulnerable, or data deficient in the 2011 IUCN Red List of Threatened Species.

	No. of Species Recently Evaluated	Recently Extinct	Critically Endangered	Endangered	Vulnerable	Data Deficient
Cetaceans	87	0	2	7	5	44
Pinnipeds	36	2	2	2	3	3
Sirenians	4	1			4	
Other carnivores (otters, polar bear)	3			2	1	

ENDANGERED

A species is endangered when it is facing a very high risk of extinction in the wild in the near future. Examples of the 12 listed endangered marine mammal species include the pinnipeds, the Galapagos fur seal (*Arctocephalus galapagoensis*) and the Steller sea lion (*Eumetopias jubatus*). More than 10 percent of cetaceans are listed in this category, including the North Pacific right whale (*Eubalaena japonica*), Sei whale (*Balaenoptera borealis*), and Hector's dolphin (*Cephalorhynchus hectori*). Also listed is the sea otter (*Enhydra lutris*) and the marine otter (*Lontra felina*).

VULNERABLE

A species is vulnerable when it faces a high risk of extinction in the wild in the medium-term future. Thirteen marine mammal species are listed in this category, including pinnipeds, the hooded seal (*Cystophora cristata*), the New Zealand sea lion (*Phocarctos hookeri*), cetaceans, the Indo-Pacific finless porpoise (*Neophocaena phocaenoides*), the narrow-ridged finless porpoise (*Neophocaena asiaorientalis*), the Irrawaddy dolphin (*Orcaella brevirostris*), sirenians, manatees (*Trichechus* spp.) and dugongs (*Dugong dugon*), and the polar bear (*Ursus maritimus*).

NEAR THREATENED

These are species that have close to vulnerable status, for which no conservation measures are in place. Seven marine mammals are listed in this category, including, of the pinnipeds, the Juan Fernandez fur seal (*Arctocephalus philippii*) and, of the cetaceans, the beluga (*Delphinapterus leucas*) and the narwhal (*Monodon monoceros*).

LEAST CONCERN

These are species that do not qualify for classification in one of the other categories. Thirty-nine marine mammal species are listed in this category, including, of the pinnipeds, the South American fur seal (*Arctocephalus australis*) and the Weddell seal (*Leptonychotes weddellii*) and, of the cetaceans, the minke whale (*Balaenoptera acutorostrata*), the bowhead (*Balaena mysticetus*), and the short-beaked common dolphin (*Delphinus delphis*).

DATA DEFICIENT

A species is listed as data deficient when information about the species' distribution and abundance is insufficient to assess its risk of extinction. The largest number of marine mammal species, 36 percent, are in this category and include, of the pinnipeds, the ribbon seal (*Histriophoca fasciata*) and the walrus (*Odobenus rosmarus*) and, of the cetaceans, the Antarctic minke whale (*Balaenoptera bonaerensis*), the pygmy right whale (*Caperea marginata*), the long-finned pilot whale (*Globicephala melas*), and the dwarf sperm whale (*Kogia simus*).

Another Approach

Some conservation biologists suggest a different approach to protecting endangered animals and plants. This new approach, vulnerability assessment, considers a broader array of information, including field observations, ecological experiments, and models incorporating ecological and evolutionary processes, to provide a more accurate picture of which species and habitats are most at risk.

Marine Mammals in Crisis and Some Solutions
COMMERCIAL WHALING AND SEALING
VERSUS ABORIGINAL HUNTING

All groups of marine mammals have been hunted at one time or another, although some have been hunted more intensively than others. The great whales (sperm whale and baleen whales) have been hunted for their oil, meat, and baleen; pinnipeds (elephant seals and northern fur seal) for their oil and pelt; sea otters for their furs; and sirenians (manatees and dugong) for their meat and skins (fig. 6.7).

Despite the fact that pelagic whaling has been banned since 1986, Japan, Iceland, and Norway and indigenous peoples killed at least 17,000 whales over the last decade. Recently, a group of countries, including the United States, has proposed lifting the moratorium on commercial whaling established by the International Whaling Commission (IWC) in 1982, permitting a limited amount of whaling. Although hunting for subsistence or survival of indigenous persons is regulated in most countries, questions have been raised concerning the use of modern hunting practices and

Figure 6.7. Canned Minke whale.

weapons by Aboriginal hunters and the fact that some products, such as seal skin and walrus ivory, are sold or traded in the global marketplace.

Marine mammals have also been hunted to reduce their consumption of valued resources such as fish or mollusks. This culling, often implemented by government-sponsored bounty programs such as that directed at harp and grey seals in Canada, continues to be controversial. Although future hunting and sealing is unlikely to approach the large scale of much of the last century, it is unsustainable for most species. Smaller whales are especially vulnerable and there are fewer regulations protecting them.

CAPTIVE BREEDING PROGRAMS

The public display of marine mammals in captivity is controversial. Proponents argue that it provides a means of generating interest and education in marine mammals and conservation. In addition, the training regimens for captive marine mammals often include behaviors designed to assist veterinarians in examining animals, which allows facilities to make more extensive use of preventive medical protocols. Arguments against captivity point to confinement of marine mammals in small pools or tanks that are not their natural habitat, and behaviors in captivity that are very different from those in the wild.

In some situations, the threat to a species is so dire and immediate that the only alternative to watching the species become extinct is to establish a captive breeding program. Unfortunately, a captive breeding program for the baiji (*Lipotes vexilifer*), which attempted to maintain this species in "seminatural reserves," was ultimately unsuccessful. Other species of marine mammals, however, are being bred in captivity with some success, including the harbor seal, California sea lion, bottlenose dolphin, and killer whale.

PROTECTED AREAS: MARINE SANCTUARIES

As large mammals, whales, pinnipeds, and sirenians need a lot of room—they have large territories and require different habitats for feeding, mating, and raising young. Critical habitat for marine mammals is protected in the United States under the 1972 Marine Protection, Research, and Sanctuaries Act. At present, there are 13 National Marine Sanctuaries and the Northwestern Hawaiian Islands Marine National Monument. Among protected critical habitats for marine mammals in the Northeast region is Stellwagen Bank, a rich feeding ground visited by more than 17 marine mammal species (for example, humpbacks, right whales, pilot whales, harbor seals) during summer months. On the West Coast, the Channel Island Marine Sanctuary serves as part of the migratory corridor for gray and blue whales. In New Zealand, Banks Peninsula Marine Sanctuary protects the rare Hector's and Maui dolphins, among other species, and in Brazil, Abrolhos National Marine Park protects humpbacks. Evaluation of the success of marine sanctuaries is an ongoing effort and has included monitoring the life history, density, and distribution patterns of various marine mammals as well as other organisms and how they relate to changes in habitat quality.

RELOCATION EXPERIMENTS—SUCCESS?

Reestablishing populations of endangered or threatened species, termed **relocation** or translocation, has become a popular conservation strategy in recent years, with the goal of increasing the number of individuals in small populations or reducing the negative effects of climate change and threat of extinction. However, relatively few projects have involved marine mammals.

In the late 1980s, the U.S. Fish and Wildlife service relocated 140 southern sea otters (*Enhydra lutris nereis*) from the mainland population near Monterey Bay, California, to San Nicholas Island. The goal was to establish a colony of sea otters that would not be affected by a nearshore catastrophic oil spill and would not deplete the commercial shellfish fishery off Southern California. Most of the otters vanished and fewer than 20 remain today. Although this translocation experiment failed, important lessons were learned. For example, the homing behavior of sea otters was not adequately considered (that is, nearly

Figure 6.8. Sea World–rescued gray whale (courtesy J. Sumich).

one-quarter of the otters returned to the capture site) nor was the high mortality rate following release of the animals. Other relocation programs have succeeded. For example, a recovery plan for the Hawaiian monk seal (*Monachus schauinslandi*), the most endangered pinniped, involved removing several female seals from Midway Island, rehabilitating them at a facility on Hawaii, and rereleasing them. This pilot study provided important information that improved the early survival of seals, and has become part of a broader strategy to conserve the species.

RECOVERY, REHABILITATION, AND RELEASE PROGRAMS

Marine mammal recovery programs are authorized in the United States under the Marine Mammal Protection Act and the Endangered Species Act. Nonprofit recovery centers, as well as public display facilities, are organized in many coastal states to rescue, rehabilitate, and release stranded marine mammals. For example, Sea World's Animal Rescue and Rehabilitation program has returned California sea lions, northern elephant seals, and harbor seals, among other species, and even two young gray whales back to the wild (fig. 6.8).

GLOSSARY

ABSOLUTE DATING The age of a rock or fossil measured in years before present.

ADAPTATION A particular biological structure, physiological process, or behavior that results from natural selection.

AEROBIC DIVE LIMIT (ADL) The length of time that an animal can remain underwater without accumulating lactic acid in the blood.

AFROTHERIA A mammalian clade with members that originate in Africa.

ALLOMETRY The variation in the relative rates of growth of various parts of the body in relation to the total animal.

ALLOMOTHERING A female that takes care of an offspring other than her own.

ALLOPATRIC SPECIATION Formation of a new species that occurs when two populations are divided by a physical barrier.

ALTRICIAL Animals that are born relatively underdeveloped.

AMBERGRIS A waxy substance found in the intestines of some whales that forms around squid beaks.

ANALOGOUS A structure that performs a similar function but has a different evolutionary origin.

ANTARCTIC CIRCUMPOLAR CURRENT An ocean current that flows from west to east around Antarctica.

ANTHROPOGENIC Caused by humans.

AQUATIC NURSING Strategy in which the young are fed at sea and mothers forage during nursing.

BACULUM Bone in the penis of some marine mammals (sea otters, polar bears, pinnipeds).

BALEEN Plates of keratin (epithelial tissue) that hang from the upper jaw of mysticete whales; also called whale bone.

BANGS or POPS Loud impulse sounds.

BENDS A serious condition that results when a diver breathes air under pressure and ascends too rapidly after spending a period of time at depth; occurs when nitrogen escapes into the blood, joints, and nerve tissue causing pain, paralysis, and death unless treated by gradual decompression.

BERING STRAIT Narrow sea passage between Asia and North America.

BIOMAGNIFICATION Series of processes in an ecosystem in which a substance increases in concentration with each higher trophic level.

BIOMIMICRY Design discipline based on inspiration and models from nature.

BLOW Exhaled air mixed with water and oil released by whales from the blowhole.

BLOWHOLE Nostril opening at the top of whale's head.

BLUBBER A thick layer of fat in the deepest layer of the skin of whales and some other marine mammals.

BOTTOM UP Refers to ecosystems controlled by nutrient availability.

BREACHING Refers to whales' leaping out of the water.

BUBBLE NET FEEDING A strategy that involves emitting a series of air bubbles underwater that form a net and serve to encircle and trap prey.

BY-CATCH That portion of a catch or harvest that includes nontargeted animals such as those taken incidentally through fishing.

CALLOSITIES Hard, thick raised skin patches on the head of right whales and a useful identification feature.

CENTRAL AMERICAN OR PANAMANIAN SEAWAY Sea that once separated North America from South America and served as a dispersal corridor for various marine mammals in the past.

CETARTIODACTYLA Monophyletic group that includes cetaceans (Cetacea) and even-toed ungulates (formerly known as Artiodactyla).

CHARACTERS Heritable morphologic, molecular, or behavioral attributes of organisms.

CLADOGRAM A branching diagram that depicts relationships among organisms based on recency of common ancestry.

CLADE A monophyletic group of organisms.

CLICKS A broad band frequency of sound of short duration.

CODAS Pattern of clicks used by sperm whales.

CONCHING A foraging behavior seen in some dolphins in which they collect conch shells and bring them to the surface of the water to flush out and eat prey.

CONVERGENT EVOLUTION A similarity among organisms that is the result of a shared ecology rather than shared ancestry.

CONTINENTAL DRIFT The idea that continents today once formed a large land mass that divided (drifted) relative to one another.

CORPUS ALBICANS A permanent visible scar on the ovary of whales that records number of ovulations though not pregnancies.

COUNTERCURRENT The opposite flow of adjacent substances that maximizes the flow of heat or energy.

COUNTERSHADING A type of camouflage in marine mammals in which the upper (dorsal) surface of the body is dark and the underside (ventral) surface is light in color.

CRITTERCAMS Cameras worn by animals.

CROWN GROUP Living members of a group and their common ancestors.

DEEP SCATTERING LAYER Concentration of zooplankton found at depth during the day that migrates to the surface at night.

DELAYED IMPLANTATION Suspension of development of the embryo in some marine mammals for several weeks or months.

DERMIS Middle layer of skin between the epidermis (outer layer) and blubber (inner layer).

DESMOSTYLIANS The only known entirely extinct higher group (order) of marine mammals.

DIATOMS A major group of algae with glass-like (silica) walls that are important primary producers.

DIGITAL ACOUSTIC TAG Small device that measures the response of wild marine mammals to sound.

DNA BAR CODING Method that uses short genetic marker (barcode) to identify an organism's DNA as belonging to a particular species.

DNA FINGERPRINTING Identification of individuals based on their DNA profile.

DOMOIC ACID A toxin produced by various marine diatom species.

DORSAL FIN Fin located on the back or dorsal surface of the body of various whales.

DRAG Refers to the force that opposes movement.

ECHOLOCATION The production of high frequency sounds and its reception by reflected echoes used by toothed whales to navigate and locate prey.

ECTOTHERM Relying on the external environment to regulate temperature.

EL NIÑO Oceanographic events that occur in the Eastern Pacific Ocean every few years involving changes in current patterns, increases in water temperature, and a decline in upwelling.

ENCEPHALIZATION QUOTIENT A numeric comparison that considers brain size relative to body size.

ENDANGERED SPECIES ACT A U.S. law passed in 1973 to establish a program to identify, protect species in danger of extinction, and provide for their recovery.

ENDOTHERM Body temperature that it is regulated by muscular activity.

EPIDERMIS Outer or top layer of skin.

EVOLUTION Descent with modification or more specifically changes in gene frequencies through time.

EXAPTATION A different function for a structure than its original function.

EXTANT Currently living as opposed to extinct.

FAMILY UNITS Social groups of females and immature sperm whales.

FASTING STRATEGY Strategy in which females rely on stored fat during lactation.

FATTY ACID SIGNATURES Technique used to examine the diet of animals based on the chemical composition of prey.

FLUKE Horizontally flattened tail of whales and sea cows.

FOOD CHAINS Sequence of who eats whom in a community.

FORAGING CYCLE STRATEGY Strategy in which females alternate between feeding at sea and caring for young.

FOSSILS Any evidence of past life including remains and traces.

FOUNDER POPULATION A few individuals from a population that colonize a new environment.

GENETIC BOTTLENECK An evolutionary event in which a significant portion of the population is prevented from reproducing.

GEOGRAPHIC LOCATION TIME DEPTH RECORDER Tracking instrument that records the location, duration, and depth of dive.

GEOLOGIC TIME SCALE Based on a hierarchy of time units (e.g. ages, epochs, periods) during which various organisms lived.

HAUL OUTS Places where pinnipeds temporarily leave the water.

HETERODONT Teeth that differ in form and function.

HOLOTYPE Single specimen designated as the type for a species.

HOMODONT Teeth that are similar in form and function to one another.

HOMOLOGOUS A similarity that is the result of inheritance.

HOX GENES Group of related developmental genes critical to the proper number and position of structures or organs (for example, legs and eyes).

HYBRIDIZATION The process of combining individuals of the same species.

HYPERPHALANGY An increase in the number of finger bones of whales.

KERPLUNKING A behavior seen is some whales that involves a loud sound usually when a heavy body or part hits the water's surface.

KEYSTONE SPECIES Species that play a critical role in an ecosystem.

KRILL Shrimp-like crustaceans that are primary food for most baleen whales.

LANUGO A coat of fine, downy hair of a newborn.

LATERAL LINE A sense organ in aquatic vertebrates used to detect movement in water.

LEK Mating strategy or courtship display in which males display to females.

LIFT Force exerted on a body that is perpendicular to the direction of flow.

LINEAGE An ancestor-descendant population.

MARINE BIODIVERSITY HOT SPOTS Restricted areas in which unique species are confined making them vulnerable to extinction.

MARINE MAMMAL PROTECTION ACT A U.S. law passed in 1972 to establish a program to manage and conserve marine mammals.

MARINE SANCTUARIES Areas of the ocean that are designated to protect marine organisms.

MASS EXTINCTION Large-scale event involving the extinction of many unrelated species.

MASS STRANDINGS Three or more individuals of the same species that intentionally swim or are unintentionally trapped ashore by waves or receding tides.

MATING HERDS Groups of males that associate with a female and attempt to mate with her.

MATRILINE Descendants of a female.

MESSINIAN SALINITY CRISIS Geologic event in the past during which the Mediterranean Sea dried up leaving large salt deposits.

MOLECULAR CLOCK Technique that uses fossil constraints and molecular rates of evolution in organisms to determine when two species diverged.

MOLT The annual shedding of a layer of skin or fur which is then replaced by new growth.

MONOGAMY Mating strategy in which males mate with only one female during a breeding season.

MONOPHYLETIC A group of organisms that includes the common ancestor and all of the descendants.

MORIBILLIVIRUSES Serious pathogens that affect marine mammals by suppressing immune system responses.

MYOGLOBIN Hemoglobin in the muscles.

NEOCETI Crown group that includes Odontocetes and Mysticetes.

NONMONOPHYLETIC A group that consists of the common ancestor but not all the descendants.

OFFSHORE KILLER WHALES One of three distinct types of killer whales consisting of populations that are reproductively isolated from other killer whale populations and spend most of their time offshore along the continental shelf.

OSMOREGULATION The control of levels of water and salt in the blood.

OSTEOSCLEROSIS Containing compact bone.

OUTGROUP Organism(s) that is closely related but outside the group whose relationships are being investigated.

PACHYOSTEOSCLEROTIC Thick, dense, compact bone.

PAEDOMORPHOSIS A change in developmental timing such that adults retain juvenile characteristics.

PARAPATRIC SPECIATION Speciation that occurs when there is reduced gene flow within a population.

PARATETHYS SEA A former large, shallow sea that extended across central Europe and eastern Asia.

PARSIMONY The principle that prefers the hypothesis with the fewest number of steps.

PERIPATRIC SPECIATION A version of allopatric speciation in which one of the isolated populations has very few individuals.

PHYLOGENETIC TREE Branching diagram that is sometimes distinguished from a cladogram as an ecological rather than a genealogical reconstruction.

PHOTOIDENTIFICATION Technique that involves the collection and use of photographs of diagnostic features of marine mammals for identification purposes.

PINGERS Acoustic alarms that are used to prevent marine mammals from net entanglement.

PINNIPEDIA Monophyletic group that includes fur seals, sea lions, and the walrus.

PLANTIGRADE Type of posture that refers to having the sole of the foot on the ground.

PLATE TECTONICS The concept that the earth's surface is organized into large mobile blocks of crustal material.

POD Social units of certain cetaceans, particularly killer whales.

POLYGYNY A mating strategy in which a male mates with more than one female in a breeding season.

PORPOISING Refers to low leaps made by some cetaceans, notably porpoises, at the surface of the water.

POSTPARTUM ESTRUS Female sexual receptivity that occurs immediately following birth.

PRECOCIAL Born in an advanced state of development.

PRIMARY PRODUCTION The initial flow of energy through an ecosystem that begins with plants converting sunlight to organic compounds.

PROMISCUOUS Mating system in which adults males randomly mate with various females.

RELATIVE DATING Determining the chronology of a sequence of rocks by the stage of evolution of the containing fossils.

RELOCATION Movement of a population of animals to a different location.

RENICULATE KIDNEY Multilobed kidney found in many marine mammals.

REPRODUCTIVE SUCCESS Passing of genes from one generation to the next.

RESIDENT KILLER WHALES Populations of killer whales with a distinctive morphology, genetics, behavior, and ecology that range from California to Russia.

RIVER DOLPHINS Four lineages of dolphins that independently evolved in fresh water.

ROSTRAL OR ORAL DISC Upper lip and mouth of manatee and dugongs covered with vibrissae that are used in feeding.

SATELLITE TELEMETRY A tracking system that uses satellites to follow the movement of animals.

SCHOOLS Structured social groups observed in odontocetes characterized by long-term associations among individuals.

SCIENTIFIC NAME Unique name of a species, consisting of genus and species names.

SECONDARY SEXUAL CHARACTERISTIC Various features that distinguish males from females.

SEXUAL BIMATURITY Males and females of species that mature at different rates.

SEXUAL DIMORPHISM External differences between males and females of a particular species.

SIRENIA Monophyletic group that includes manatees and dugongs.

SISTER SPECIES A species pair that are each other's close relatives.

SPECIATION Splitting of lineages resulting in the formation of two species.

SPERMACETI Waxy fluid that fills the spermaceti organ and was once sought after by hunting whales.

SPERM COMPETITION Mating strategy seen in some mysticetes in which copulation males attempt to displace or dilute the sperm of other males in an attempt to increase the probability of being the male to fertilize that female.

SPONGING A behavior seen in some dolphins in which they use sponges to flush out prey.

SPY HOPPING Behavior in which a whale comes partially or entirely out of the water.

STABLE ISOTOPES Forms of various chemical elements that on the basis of their mass:change ratio can be used to reconstruct diet.

STEM GROUP Organisms that fall outside the crown group.

SURFACE AREA: VOLUME Ratio that is the amount of surface area per unit volume of a object or body.

SYMPATRIC SPECIATION Type of speciation in which a new species evolves in the same area by reduced gene flow as may occur when some individuals exploit a new niche.

SYNAPOMORPHY A shared derived character that provides evidence of a shared common ancestry.

TAPHONOMY Discipline that deals with various processes that take place from when an organism dies to when it is recovered.

TAXONOMY A branch of systematics that deals with the identification, description, and classification of species.

TELESCOPED SKULL Changes in the relationships of bones of the whale skull in response to migration of the nasal opening to the top of the skull.

TETHYS SEA A former shallow sea that extended in areas that are now occupied by India and Pakistan.

TETHYTHERIA Monophyletic group that includes proboscideans (elephants), sirenians, and extinct desmostylians.

THERMOREGULATION Ability of an organism to regulate its temperature.

THRUST Force which moves an organism in a particular direction.

TOP DOWN Refers to ecosystem controlled by predators.

TRANSIENT KILLER WHALES Populations of killer whales with a distinct morphology, genetics, behavior, and ecology (feeding almost exclusively on marine mammals) that range along the west coast of North America.

TROPHIC LEVEL The division of species in an ecologic community based on their main food source.

UNDERFUR Short, fine hairs that function primarily in insulation.

UPWELLING Area off continental margins where circulation patterns bring nutrient-rich water to the surface.

VIBRISSAE Whiskers.

WAVE OR BOW RIDING Behavior seen in dolphins in which they ride the waves of a boat's wake.

WEANER PODS Groups of weaned elephant seal pups.

WHALE-FALL COMMUNITIES Organisms that evolve around the carcasses of whales on the sea bottom.

WHALE LICE Amphipod crustaceans (for example, cyamids) found on the skin of some large whales.

WHISTLES Narrow frequency band sounds produced by some dolphins for individual recognition, also known as signature whistles.

WORLD CONSERVATION UNION Comprehensive organization to sustainably manage and conserve the world's biodiversity.

WORLD HERITAGE SITE Area Listed by the United Nations Educational, Scientific, and Cultural Organization (UNESCO) as a special place of cultural or physical significance.

FURTHER READING AND ONLINE SOURCES

CHAPTER 1: MARINE MAMMALS

Further Reading

Baker, C. S., M. L. Dalebout, S. Lavery, and H. A. Ross. 2003. "DNA-Surveillance: Applied Molecular Taxonomy for Species Conservation and Discovery." *Trends in Ecology and Evolution* 18 (6): 271–72.

Berta, A., J. L. Sumich, and K. M. Kovacs. 2006. *Marine Mammals: Evolutionary Biology.* 2nd ed. San Diego, CA: Elsevier (Academic Press).

Perrin, W. F., B. Wursig, and J. G. M. Thewissen. 2009. *The Encyclopedia of Marine Mammals.* 2nd ed. San Diego, CA: Academic Press.

Online Sources

Encyclopedia of Life (http://www.eol.org)
  A project to provide information about biodiversity knowledge about all known species, including their taxonomy, geographic distribution, collections, genetics, evolutionary history, morphology, behavior, and ecological relationships.

Overwinter Movement Patterns of Antarctic Predators
  (http://swfsc.noaa.gov/AntarcticPredators)
  National Oceanic and Atmospheric Administration (NOAA) scientists track penguins, fur seals, and seals in Antarctica.

Tagging of Pacific Predators (http://www.topp.org)
  An international team of scientists track 22 species of predators in the

Pacific, including sharks, turtles, elephant seals, and birds, providing valuable information on the migration routes and foraging activities.
Understanding Evolution (http://evolution.berkeley.edu)
A comprehensive overview of evolution, with resources.

CHAPTER 2: PAST DIVERSITY IN TIME AND SPACE, PALEOCLIMATES, AND PALEOECOLOGY

Further Reading

Berger, W. H. 2007. "Cenozoic Cooling, Antarctic Nutrient Pump, and the Evolution of Whales." *Deep-Sea Research II* 54: 2399–421.

Deméré, T. A., A. Berta, and P. J. Adam. 2003. "Pinnipedimorph Evolutionary Biogeography." *Bulletin of the American Museum of Natural History* 279: 32–76.

Ellis, R. 1994. *Monsters of the Deep.* New York: Doubleday.

Smith, A. G., D. G. Smith, and B. M. Funnel. 1994. *Atlas of Mesozoic and Cenozoic Coastlines.* Cambridge, UK: Cambridge University Press.

Wallace, D. R. 2007. *Neptune's Ark from Ichthyosaurs to Orcas.* Berkeley, CA: University of California Press.

Online Sources

Earth History (http://www.scotese.com)
A project to illustrate development of continents and ocean basins.
Plate Tectonics (http://serc.carleton.edu/NAGTWorkshops/geophysics/visualizations/-reconstructions)
Shows plate movement through time.

CHAPTER 3: PINNIPED DIVERSITY, EVOLUTION, AND ADAPTATIONS

Further Reading

Bonner, N. 1994. *Seals and Sea Lions of the World.* New York: Facts on File.

Fay, F. H. 1982. "Ecology and Biology of the Pacific Walrus." *Odobenus rosmarus divergens* Illiger. U.S. Department of the Interior Fish and Wildlife Services, North American Fauna. No. 74.

Gentry, R. 1998. *Behavior and Ecology of the Northern Fur Seal.* Princeton, NJ: Princeton University Press.

Gentry, R., and G. L. Kooyman. 1986. *Fur Seals: Maternal Strategies on Land and Sea.* Princeton, NJ: Princeton University Press.

King, J. E. 1982. *Seals of the World.* 2nd ed. London: British Museum of Natural History.

Kooyman, G. L. 1981. *Weddell Seal: Consumate Diver.* Cambridge, MA: Cambridge University Press.

Lavigne, D. M., and K. M. Kovacs. 1988. *Harps and Hoods.* Ontario: University of Waterloo Press.

Le Boeuf, B. J., and R. M. Laws, ed. 1994. *Elephant Seals.* Berkeley, CA: University of California Press.

Riedman, M. L. 1990. *The Pinnipeds.* Berkeley, CA: University of California Press.

Rybczynski, N., M. R. Dawson, and R. H. Tedford. 2009. "A Semi-Aquatic Arctic Mammalian Carnivore from the Miocene Epoch and Origin of Pinnipedia." *Nature* 458 (7241): 1021–24.

Williams, T. M. 2004. *The Hunter's Breath: On Expedition with the Weddell Seals of the Antarctic.* New York: M. Evans and Company.

Online Sources

Long Marine Lab, University of California, Santa Cruz (http://www.pinnipedlab.org)
Pinniped research in cognition and sensory systems.

Seal Conservation Society (http://www.pinnipeds.org)
An international organization providing information about pinnipeds and their conservation.

CHAPTER 4: CETARTIODACTYLAN DIVERSITY, EVOLUTION, AND ADAPTATIONS

Further Reading

Carwardine, M. 1995. *Whales Dolphins and Porpoises.* New York: Dorling Kindserley Publishing.

Ellis, R. 2006. *Singing Whales and Flying Squid. The Discovery of Marine Life.* Guilford, CT: Lyons Press.

Ellis, R. 1985. *The Book of Whales.* New York: Knopf.

Ford, J. B., G. M. Ellis, and K. C. Balcomb. 2000. *Killer Whales: The Natural*

History and Genealogy of Orcinus orca in British Columbia and Washington State. Seattle, WA: University of Washington Press.

Gingerich, P. D. 2009. "New Protocetid Whale from the Middle Eocene of Pakistan: Birth on Land, Precocial Development, and Sexual Dimorphism." *PLoS One* 4 (2): e4366. doi: 10.1371/journal.pone.0004366

Goldbogen, J. A. 2010. "The Ultimate Mouthful: Lunge Feeding in Rorqual Whales." *American Scientist* 98: 124–31.

Jones, M. L., S. Schwartz, and S. Leatherwood. 1984. *The Gray Whale.* New York: Academic Press.

Kelsey, E. 2009. *Watching Giants.* Berkeley, CA: University of California Press.

Lindberg, D. R., and N. D. Pyenson. 2007. "Things That Go Bump in the Night: Evolutionary Interactions between Cephalopods and Cetaceans in the Tertiary." *Lethaia* 40: 335–43.

Mann, J., R. C. Connor, P. L. Tyack, and H. Whitehead. 2000. *Cetacean Societies: Field Studies of Dolphins and Whales.* Chicago, IL: University of Chicago Press.

Marx, F. G., and M. D. Uhen. 2010. "Climate, Critters, and Cetaceans: Cenozoic Drivers of the Evolution of Modern Whales." *Science* 327: 993–5.

Norris, K., B. Wursig, R. S. Wells, and M. Wursig, ed. 1994. *The Hawaiian Spinner Dolphin.* Berkeley, CA: University of California Press.

Reynolds, J. E. III, R. S. Wells, and S. D. Eide. 2000. *The Bottlenose Dolphin: Biology and Conservation.* Gainesville, FL: University of Florida Press.

Thewissen, J. G. M., L. N. Cooper, J. C. George, and S. Bajpal. 2009. "From Land to Water: the Origin of Whales, Dolphins, and Porpoises." *Evolution, Education Outreach* 2: 272–88.

Whitehead, H. 2003. *Sperm Whales: Social Evolution in the Ocean.* Chicago: University of Chicago Press.

Zimmer, C. 1998. *At the Water's Edge.* New York: Free Press.

Online Sources

American Cetacean Society (http://www.acsonline.org)
    Whale and dolphin conservation, education, and research.

European Cetacean Society (http://www.EuropeanCetaceanSociety.eu)
    An organization to promote and advance study and conservation of cetaceans and other marine mammals.

WhaleNet (http://whale.wheelock.edu/WhaleNet)
    An interactive educational website focused on whales.

CHAPTER 5: DIVERSITY, EVOLUTION, AND
ADAPTATIONS OF SIRENIANS AND
OTHER MARINE MAMMALS

Further Reading

Domning, D. P. 2001. "The Earliest Known Fully Quadrupedal Sirenian." *Nature* 413: 625–7.

Harmon, K. 2010, March. "Climate Change Likely Caused Polar Bear to Evolve Quickly." *Scientific American.*

Muizon, C. de, H. G. McDonald, R. Sala, and M. Urbina. 2004. "The Youngest Species of the Aquatic Sloth and a Reassessment of the Relationships of the Nothrothere Sloths (Mammalia: Xenarthra)." *J. Vert. Paleo* 24: 387–97.

Reep, R. L., and R. K. Bonde. 2006. *The Florida Manatee Biology and Conservation.* Gainesville, FL: University of Florida Press.

Reynolds, J. E. III, and D. K. Odell. 1991. *Manatees and Dugongs.* New York: Facts on File.

Riedman, M. L. 1990. *Sea Otters.* Monterey, CA: Monterey Bay Aquarium Foundation.

Stirling, I. 1999. *Polar Bears.* Ann Arbor, MI: University of Michigan Press.

Turvey, S. T., and C. L. Risley. 2006. "Modelling the Extinction of Steller's Sea Cow." *Biology Letters* 2: 94–97.

Online Sources

Save the Manatee Club (http://www.savethemanatee.org)
Provides information about manatees, especially conservation efforts.

Saving Sea Otters (http://www.montereybayaquarium.org)
Otter research and conservation at Monterey Bay Aquarium, California.

Sirenian International (http://www.sirenian.org)
Manatee and dugong research, education, and conservation.

West Coast Sea Otter Recovery (http://www.oceanlink.info/seaotterstewardship)
History, biology, and conservation of sea otters, Bamfield Marine Center, British Columbia, Canada.

CHAPTER 6: ECOLOGY AND CONSERVATION

Further Reading

Ellis, R.1991. *Men and Whales.* New York: Knopf.

Ellis, R. 2009. *On Thin Ice: The Changing World of the Polar Bear.* New York: Knopf.

Estes, J. A., D. P. DeMaster, D. F. Doak, and T. M. Williams. 2007. *Whales, Whaling, and Ocean Ecosystems.* Berkeley, CA: University of California Press.

Jackson, J. B. C., M. K. Kirby, W. H. Berger, et al. 2001. "Historical Overfishing and the Recent Collapse of Coastal Ecosystems." *Science* 293: 629–38.

Pyenson, N. D., and D. R. Lindberg. 2011. "What Happened to Gray Whales During the Pleistocene? The Ecological Impact of Sea-Level Change on Benthic Feeding Areas in the North Pacific." *PLoS ONE* 6 (7): e21295. doi: 10.1371/journal.pone.0021295

Pittman, C. 2010. *Manatee Insanity: Inside the War over Florida's Most Famous Endangered Species.* Gainesville, FL: University of Florida Press.

Reeves, R. R., and J. R. Twiss. 1999. *Conservation and Management of Marine Mammals.* Washington, DC: Smithsonian Institution Press.

Reynolds, J. E. III, W. F. Perrin, R. R. Reeves, S. Montgomery, and T. J. Ragan, ed. 2005. *Marine Mammal Research Conservation beyond Crisis.* Baltimore, MD: Johns Hopkins University Press.

Online Sources

International Whaling Commission (http://www.iwcoffice.org)
    An organization for conservation and management of whales and whaling.

Marine Mammal Center (http://www.marinemammalcenter.org)
    The center is involved in the rescue, rehabilitation, and release of marine mammals.

Marine Protected Area (http://www.cetaceanhabitiat.org)
    News and directory of global cetacean protected habitats.

National Marine Sanctuaries (http://sanctuaries.noaa.gov)
    Information about U.S. marine sanctuaries.

ILLUSTRATION CREDITS

Figs. 1.3, 1.8, 1.9. 1.10, 1.11, 3.8, 3.11, 3.12, 3.13, 3.14, 3.15, 4.4, 4.20, 4.22, 4.23, 4.26, 5.2, 5.11: Berta, A., J. L. Sumich, and K. M. Kovacs. 2006. *Marine Mammals: Evolutionary Biology.* 2nd ed. San Diego, CA: Elsevier (Academic Press).

Fig. 1.12: Costa, D. P., P. W. Robinson, J. P. Y. Arnould, A. L. Harrison, S. E. Simmons, J. L. Hasrick, A. J. Hoskins, S. P. Kirkman, H. Osthuizan, S. Villegan-Antmann, and D. E. Crocker. 2010. "Accuracy of Argos Locations of Pinnipeds at Sea Estimated Using Fastloc GPS." *PloS One* 5 (1): E8677. doi: 10.1371/Journal.Pone.0008677

Fig. 1.13: Watwood, S. et al. 2006. "Deep-Diving Foraging Behaviour of Sperm Whales (*Physeter macrocephalus*)." *Journal of Animal Ecology* 75: 814–25.

Fig. 2.2: Heuvelmans, B. 1968. *In the Wake of the Sea Serpents.* New York: Hart-Davis.

Figs. 2.3, 2.6, 2.8: Berta, A., J. L. Sumich, and K. M. Kovacs. 2006. *Marine Mammals: Evolutionary Biology.* 2nd ed. San Diego, CA: Elsevier (Academic Press) and Fordyce, R. E. 2008. "Fossil Sites." In *The Encyclopedia Of Marine Mammals*, 2nd ed., edited by W. F. Perrin, B. Wursig, and J. G. M. Thewissen, 459–66. San Diego, CA: Academic Press.

Fig. 2.4: Marx, F., and M. Uhen. 2010. "Climate, Critters, and Cetaceans: Cenozoic Drivers of the Evolution of Modern Whales." *Science* 327: 993–96.

Fig. 2.5: Uhen, M. D. 2007. "Evolution of Marine Mammals: Back to the Sea after 300 Million Years." *The Anatomical Record* 290: 514–22.

Fig. 2.12: Fordyce, R. E. 2008. "Fossil Sites." In *The Encyclopedia Of Marine Mammals*, 2nd ed., edited by W. F. Perrin, B. Wursig, and J. G. M. Thewissen, 459–66. San Diego, CA: Academic Press.

Fig. 4.8: Lambert, O., G. Bianucci, K. Post, C. De Muizon, R. Salas-Gismondi, M. Urbina, and J. Reumer. 2010. "The Giant Bite of a New Raptorial Sperm Whale from the Miocene Epoch of Peru." *Nature* 466: 105–8.

Fig. 4.11: Muizon, C. de. 1993a. "Walrus Feeding Adaptation in a New Cetacean from the Pliocene of Peru." *Nature* 365: 745–48 and Muizon, C. de. 1993b. "*Odobenocetops peruvianus*: Una Remarkable Convergenciade Adaptacion Alimentaria Entre Morsa y Dolphin." *Bull. d'l'Institutut Francais d'Etudes Andines* 22: 671–83.

Fig. 4.18: Lindberg, D. R., and N. D. Pyenson. 2007. "Things that Go Bump in the Night: Evolutionary Interactions between Cephalopods and Cetaceans in the Tertiary." *Lethaia* 40: 335–43.

Fig. 4.19: Friedman, M., K. Shimada, L. D. Martin, M. J. Everhart, J. Liston, A. Maltese, and M. Triebold. 2010. "100-Million-Year Dynasty of Giant Planktivorous Bony Fishes in the Mesozoic Seas." *Science* 327: 990–93.

Fig. 4.21: Fish, F. E., and J. M. Battle. 1995. "Hydrodynamic Design of the Humpback Whale Flipper." *Journal of Morphology* 225: 51–60.

Fig 4.24: Cranford, T. W., M. Amundin, and K. S. Norris. 1996. "Functional Morphology and Homology in the Odontocete Nasal Complex: Implication for Sound Generation." *J. Morphol* 228: 223–85.

Fig 4.27: Goldbogen, J. A. 2010. "The Ultimate Mouthful: Lunge Feeding in Rorqual Whales." *American Scientist* 98: 124–31.

Fig. 5.4: Hartman, D. S. 1979. "Ecology and Behavior of the Manatee (*Trichechus manatus*) in Florida." *American Society of Mammalogists Special Paper* 5: 1–153.

Fig. 5.8: Marsh, H., C. A. Beck, and T. Vargo. 1999. "Comparison of the Capabilities of Dugongs and West Indian Manatees to Masticate Seagrasses." *Mar. Mamm. Sci.* 15 (1): 250–55.

Fig. 5.9: Reynolds, J. E. III, and S. A. Rommel. 1996. "Structure and Function of the Gastrointestinal Tract of the Florida Manatee, *Trichechus manatus latirostris*." *Anatomical Record* 245: 539–58.

Fig. 5.13: Estes, J. A., and J. L. Bodkin. 2002. "Otters." In *The Encyclopedia of Marine Mammals*, 1st ed., edited by W. F. Perrin, B. Wursig, and J. G. M. Thewisson, 845. San Diego, CA: Academic Press.

Fig. 6.2: Estes, J. A., M. T. Tinker, T. M. Williams, and D. F. Doak. 1998. "Killer Whale Predation on Sea Otter Linking Oceanic and Nearshore Ecosystems." *Science* 282: 473–76.

Fig. 6.4: Roman, J., and J. J. McCarty. 2010. "The Whale Pump: Marine Mammals Enhance Primary Productivity in a Coastal Basin." *PLoS One* 5 (10).

Fig. 6.5: National Research Council. 1992. *Dolphins and the Tuna Industry.* Washington, DC: National Academy Press.

Fig. 6.6: Malakoff, D. 2010. "A Push for Quieter Ships." *Science* 328: 1502–3.

INDEX

Note: Page numbers with *f* indicate figures; those with *t* indicate tables.

Text:	10.75/15 Janson MT Pro
Display:	Janson
Compositor:	Scribe Inc.
Printer and binder:	Sheridan